Pra

"Morris a[...] like an epidemiologist, and the people and events behind waterborne disasters like an investigative reporter. . . . The effect is riveting. . . . [*The Blue Death*] makes a convincing case that more attention must be paid to our water supply."

—*Kirkus Reviews* (starred review)

"It's not every day that you read a history of microbiology and waterborne illnesses and walk away thinking you've read a detective novel, but that's exactly what Morris has achieved with *The Blue Death*, an informative work of nonfiction that entertains as well."

—*Lexington Herald-Leader*

"While casual readers don't generally pick up public health books expecting to stay up late turning pages, Morris manages a neat trick—he provides an in-depth medical history that at times reads like a mystery."

—*San Francisco Chronicle Book Review*

"Be afraid. As a topic for a summer page-turner, you'd think municipal water safety would be a tad dry. Think again. Starting with cholera's invasion of London in 1831, Morris shows that when it comes to waterborne threats, *Jaws*'s great white has nothing on bacteria and protozoa, which still stalk contemporary sewers, aqueducts, and city filtration plants. You'll never drink water again!"

—*Wired*

"Morris recounts, with crystal clarity, some of history's epic drinking water disasters. . . . [He] put the words *death*, *disease*, and *disaster* in the book's title to warn readers that his no-holds-barred narrative isn't for the squeamish. Pass the vodka, please. Uh, no ice."

—*Booklist* (starred review)

ABOUT THE AUTHOR.

DR. ROBERT D. MORRIS is an internationally recognized expert in the field of drinking water and health. In addition to being a physician, he holds a Ph.D. in environmental engineering and an MS degree in biostatistics and epidemiology. Reports of his research have been featured in media outlets throughout the world, including *Dateline NBC*, the *New York Times*, the BBC, and *The Times* of London. This is his first book.

THE BLUE DEATH

THE INTRIGUING PAST AND PRESENT DANGER OF THE WATER YOU DRINK

Dr. Robert D. Morris

HARPER

NEW YORK · LONDON · TORONTO · SYDNEY

To Astrid, Hana, Darwin, Sage, and Skyler,
for the joy you bring to life

HARPER

A hardcover edition of this book was published in 2007 by HarperCollins
Publishers.

HarperCollins books may be purchased for educational, business, or sales
promotional use. For information please write: Special Markets Department,
HarperCollins Publishers, 10 East 53rd Street, New York, NY 10022.

FIRST HARPER PAPERBACK PUBLISHED 2008.

DESIGNED BY MARY AUSTIN SPEAKER

The Library of Congress has catalogued the hardcover edition as follows:
  Morris, Robert D.
    The blue death: disease, disaster, and the water we drink / Robert
      Morris.—1st ed.
    p. cm.
    Includes bibliographical references. ·
    ISBN: 978-0-06-073089-5
    ISBN-10: 0-06-073089-7
    1. Waterborne infection—history. I. Title.
  RA642.W3 M67 2007
  614.4'3—dc22   2006049674

ISBN 978-0-06-073090-1 (pbk.)

08 09 10 11 12 ID/RRD 10 9 8 7 6 5 4 3 2 1

CONTENTS

ILLUSTRATIONS

Dr. John Snow, 1813–1858. A photograph in the series Literary and Scientific Portrait Club. (Bowerbank J. S. Literary and Scientific Portrait Club. London: Heinz Archive and Library of the National Portrait Gallery.)

Portrait of Sir Edwin Chadwick. (Photograph by J. Delmege. Courtesy of the Wellcome Library, London.)

Robert Koch, 1843–1910, bacteriologist, working in his laboratory at Kimberley, South Africa, 1921. (Photograph by William Osler.)

Vibrio cholerae. (Photograph from *Arbeiten aus dem Kaiserlichen Gesundheitsamte,* Volume III, 1887. Julius Springer Verlag, Berlin.)

Cryptosporidium. (Photograph courtesy of Dr. Saul Tzipori, Tufts University School of Veterinary Medicine.)

Carrollton Water Treatment Plant. (Photograph by Dr. Robert D. Morris, 2005.)

Collecting water in Mali. (Photograph by João Silva.)

PROLOGUE

Drinking water. In the walls, beneath the streets, around the world, it races through unseen pipes to fill tens of billions of glasses, cups, and bottles each day and to quench that most essential of human drives, thirst. For millions of years, intimate knowledge about the source of our water was among the most important bits of information our ancestors carried. Today that intimacy is lost. We turn on a tap and water flows as if by magic. We have come to accept the illusion as reality. Most of us have little awareness of the source of our drinking water. We assume it will be there. We assume it will be safe.

The road to disaster is paved with assumptions. The largest waterborne outbreak in U.S. history happened not centuries ago, but in 1993. Not only does waterborne disease still happen, but we don't even know how often it occurs. Our system for detecting waterborne disease is so limited that drinking water is never even recognized as the cause in the vast majority of cases. Evidence suggests that drinking water may sicken millions of people every year in the United States.

For much of the developing world, waterborne disease is no secret. Like a tsunami in slow motion, unsafe drinking water is killing constantly; almost forty thousand people will die this week alone. Unlike a tsunami, it never stops.

In 1994 cholera swept through a crowded refugee camp in Goma, Zaire, and killed sixty thousand people in less than a month. It was the worst outbreak of waterborne disease in human history. The horror of Goma lies so far beyond the realm of experience for most of us that it takes on a sense of the remote and abstract. The gap between an epidemic in Goma and the sanitary comfort of the developed world seems vast, but for many reasons, this chasm may not be as immense as we imagine. Just a hundred years ago, waterborne typhoid fever was a leading cause of death in the United States. Less than fifty years before that, the major cities of Europe and North America were ravaged by waterborne cholera. The only thing that separates us from Goma is the systems we have developed to transport and treat our sewage and drinking water.

The operation of our water supplies is, to most of us, invisible. Invisibility encourages complacency. We have come to think of these systems as failsafe, but the technology we rely on for treating most of our drinking water is almost a century old and many of our water treatment plants have been in operation since the early twentieth century.

At least some of the water from these aging plants is, quite literally, treated sewage. Farm runoff, industrial waste, and sewage, both treated and untreated, routinely find their way to the intakes of our water treatment plants. Studies have shown that some of the pathogens (disease-causing microbes) from these sources can and do make their way into drinking water supplies, sometimes causing devastating outbreaks and frequently causing sporadic cases of disease. These diseases are not as deadly as cholera, but it is possible that this may not always hold true.

To understand where this story might lead, we must turn to its beginning. We must go back to a time when the difference between Goma and the developed world was far smaller, a time when we understood far less about health and disease and had no idea that a glass of water could kill.

In 1827, the industrial revolution was redefining the cities of the world. These population centers had grown over centuries from their agrarian roots into centers for commerce, education, religion, and government. Then, in a matter of decades, they had become the foundation of an uncharted industrial future, but remained propped on a rickety, haphazard infrastructure. Ill equipped to handle the influx of workers and the excreta of industry, these cities were straining at the seams. Filth and squalor grew in lockstep with urban populations. On the back of squalor rode epidemic diseases. When that happened the remarkably backward world of eighteenth century medicine would find itself scrambling to understand the causes of these diseases to identify the mechanisms for their control. At stake was nothing less than the viability of the industrial city.

Dr. John Snow   Sir Edwin Chadwick

PART I

Waterborne Killers

"Look at the water. Smell it! That's wot
we drinks. How do you like it, and what
do you think of gin instead!"

CHARLES DICKENS, *Bleak House*

THE BLUE DEATH

As John Snow stood on the streets of York and bid farewell to his father, the air swirled with traces of spring, the odor of horses, and the ever-present reminders of bad sanitation. He climbed aboard the waiting coach with the few items of clothing that his father's meager income could provide, food that his mother had prepared earlier that day, and the improbable hopes of his parents.

The crack of the driver's whip bisected the life of young John Snow. His childhood dissolved into memories as the carriage rattled off the cobblestones of York to the ringing beat of horses' hooves. As he bounced north along the turnpike to Newcastle, his future began.

In time John Snow would reshape medical science, invent the fundamental tools of epidemiology, and redefine our relationship with drinking water. But in that moment, he was just a fourteen-year-old boy, alone in the shadows of the carriage. Through its window, he watched the landscape of the familiar disappear. The year 1827 offered no time for the indulgence of adolescence. He would not see his parents again for seven years.

Snow had come of age amid the poverty that hugged the banks of the River Ouse. As the son of a laborer, he might well have expected to spend his life in a hardscrabble neighborhood like the one into which he had been born. The river brought ships and barges and the opportunity for work, but it was grueling, physical labor that could grind a man to the bone with little chance for advancement. All manner of vermin, human as well as animal, scurried along the riverside. For a child, danger lurked in every darkened corner of the district.

One of the greatest hazards was the river itself. It routinely overflowed its banks, leaving behind dankness and rot. When it stayed within its course, many of the Snows' neighbors along North Street routinely drank its water, oblivious to the hazards it carried.

John's chances of escaping the filth and disease that clung to the working poor were slim. If the daunting financial, physical, and social realities were not enough, Fanny Snow, the illegitimate daughter of a Yorkshire weaver, was heavy with her eighth child when she put her oldest son on that carriage to Newcastle. The simple demand of supporting such a large family would seem to extinguish any hope of escaping their place at the bottom of the economic ladder. The Snows, however, were not an average working class couple and John was far from a typical son.

The journey to Newcastle began when a six-year-old boy walked along Far Water Lane, turned down a narrow alley, and, for the first time, entered a remarkable world. There in the single room that comprised the Dodsworth School in St. Mary's Parish, John Snow's insatiable drive to understand took root. John Dodsworth, a York ironmonger, had founded three such schools to offer education to the city's poor. The school Snow attended offered only twenty spots for boys between the ages of six and fourteen, selecting only the most talented and deserving children. With three parishes vying for just three or four openings each year, John may well have been the only child from the parish of All Saints Church chosen that year to attend. At Dodsworth School, he could learn to read and write free of charge. Arithmetic, his favorite subject, cost extra.

This was a fortuitous beginning for the bright young boy. For the eight years he attended, his parents not only made do without the assistance of their son, but also scraped together the extra money for his foray into math and science. Once he had completed those early years of schooling, he was ready to take a remarkable next step. John Snow would become a doctor.

The carriage rattled north across the English countryside for twenty-one bone-jarring hours before John Snow rolled through Gateshead, crossed the River Tyne, and rode into Newcastle. The view out the carriage window was unlike anything he had ever seen. The young man from York stared out at the grand metropolis. Great sailing ships lined the river, waiting to carry away the coal that powered the engines of the world and the booming economy of Newcastle. Ahead, on a hill, the castle keep stood watch over the bustling city as the spires of St. Nicholas and All Saints Church pierced the industrial sky.

The carriage left him in the heart of the city. From there John Snow walked up Westgate Street in the shadow of the thick stone tower of St. John's church. There on the hillside, far from the filth and stink of the river's edge, lived the city's well-to-do. He had never seen such fine houses. Now he would live in one. For the next four years, he would stay in the home of William Hardcastle, just across from the church. A surgeon apothecary who had begun his practice in York before moving to Newcastle, Hardcastle was now among the most prominent doctors in the city. For a fee of one hundred guineas, he had agreed to take on Snow as an apprentice.

It seems likely that a hidden hand nudged open the door of opportunity to admit John Snow. The apprenticeship fee alone, roughly thirteen thousand in today's dollars, would have dissuaded even the hardest-working laborer in 1827. Even with the fee in hand, it seems unlikely that an established surgeon would have taken on a poor boy from York as an apprentice. But more than five thousand miles away, in the jungles of South America, John Snow had a friend.

For three years Charles Empson had traveled deep into the Andean rain forest riding mules and small boats hundreds of miles into what would become Colombia. He had braved snakes, poisonous insects, and well-armed thieves and had dined on everything from roast armadillo to tortoise hash. He had come with the engineer Robert Stephenson to search the region's abandoned gold and silver mines for business opportunities.

Empson was the brother of Fanny Snow. Although he would one day become a man of means, he was not yet wealthy. But he already had something far more valuable than mere money. Charles Empson had connections. He possessed a charisma that could unlock the doors of British society, which allowed him to create an ever-expanding social network. George Stephenson, the father of his traveling partner, was a visionary pioneer in the development of the British railroads. George and Robert Stephenson would go on to establish the first company to manufacture locomotives in England. The Stephensons lived just outside of Newcastle. Their family physician was another of Empson's closest friends and John Snow's mentor, William Hardcastle.

Empson and Snow would share a remarkable lifelong intimacy. Today the two men lie buried side-by-side in Brompton Cemetery. In 1827 Empson had connections to money, Newcastle, and Hardcastle, and he knew that his nephew, with his remarkable aptitude for math and natural science, was preparing to begin his career. It is not clear exactly which strings Empson pulled as he sat in Bogota and planned his return to England, but it is almost certain he pulled them.

If industry, ability, and a benevolent uncle launched John Snow on this mission, fate would define its course. A world away, in the ghettoes of Calcutta, another journey was beginning. An epidemic like none the world had seen before had begun to spread. As John Snow arrived to begin his medical training in Newcastle, cholera was in India, packing its bags.

In 1827 as William Hardcastle introduced his young apprentice to the vagaries of nineteenth-century surgery, cholera began to stretch its first fingers of death to the north and west of Calcutta. As John Snow learned the proper ways to slice into a patient's veins and drain his blood, the disease climbed into the rugged mountains of central Asia.

Cholera strikes quickly. Within a day or two, its victims are writhing, immobilized in its terrifying grip. An obligate parasite, it must depend on its unfortunate host for survival. In that host, the bacteria finds food and an ideal environment to reproduce. The poor man or woman must also help cholera find a new victim.

Travel through the mountains of Afghanistan and southern Russia was arduous even under the best conditions. In the cold of winter, cholera's messengers slowed to a crawl. The mountains were almost unpopulated. The carriers died or recovered before they could find new victims and cholera's advance on Europe stalled.

Cholera, however, is a patient killer. By 1829 it had a second chance to spread. Improvements in the routes of transportation and the steady flow of British and Russian troops allowed cholera to reach Moscow. The great powers resorted to desperate measures to halt the disease. The Russians ordered their armies to surround any town where cholera appeared and shoot those who sought to escape. As the disease moved westward, the Germans massed troops at their border in hope of stopping the advance of the epidemic. Military might, however, was no match for cholera. By 1831, four years after John Snow began his apprenticeship, the British were under siege.

A creature that comes to life in the warmth of the human gut, the cholera bacterium thrives in hot, humid environments. Away from its home in India, it advanced in the summer and hid in the winter. During the summer of 1831, the pathogen took control of the ports of continental Europe. Britain's vast armada of merchant ships flowed steadily into and out of those contaminated harbors. Each new load of returning cargo threatened to bring death to England.

Late in the summer, the Privy Council in London mandated that ships from Russia, Germany, or any Baltic port sit in quarantine for fifteen days. British warships patrolled the harbors of England, their cannons keeping watch over the invisible threat. As summer gave way to fall, the quarantine appeared to be working. By October the heat was fading and with it, the chances that a cholera epidemic would take hold. But Britain would not be spared. In the port city of Sunderland, at the mouth of the River Wear, the defenses of the realm were unraveling.

The River Wear was William Sproat's life. The river ran just south of Newcastle and the region's burgeoning coal industry had been good for business. The robust keelman spent most of the fall plying the river in heavily laden barges. Occasionally he would reach a thirsty hand into its dark water. It seems that one day in the fall of 1829, as he drew his hand back from the river that had, for so long, given him life, death clung to his fingers.

In the middle of October, something took hold of William Sproat. He fought his illness for more than a week. Disease was a constant companion in nineteenth-century England. Sproat had seen all the illnesses of the day, but none had proved a match for his sturdy frame. He had never felt anything like this.

During the dark, early hours of Sunday, October 23, the disease got the upper hand. After ten days of vomiting and violent diarrhea, excruciating cramps wracked William Sproat's body. The family doctor had no remedy. Fearing the worst, Sproat's wife rushed to the home of the one physician who might offer hope.

Dr. James Butler Kell, the only doctor in Sunderland who had ever seen a case of cholera, was surprised to find the desperate Mrs. Sproat at his doorstep. Kell, an army surgeon, had recently come to Sunderland after twenty-eight years of military service that had taken him to the far reaches of the British Empire and into cholera's kingdom. When Mrs. Sproat pleaded for his help, he pointed out that she had a physician already and he did not want to intrude on the practice of a local doctor. As she continued to describe the state of her husband, Kell's memory stirred. The more he learned, the more convinced he grew that he should examine Mr. Sproat. Nonetheless he did not want to do so on his own. He immediately sent an urgent message to Dr. Reid Clanny, the most respected physician in Sunderland and a member of the newly formed Sunderland Board of Health, requesting that he join him in visiting the afflicted boatman.

Within an hour, Dr. Kell and Dr. Clanny entered the home of the Sproats, a comfortable house on Fish Quay overlooking the harbor. Fearing the worst, William Sproat's family, including his adult son and his eleven-year-old granddaughter, waited anxiously as the two doctors attended to their gravely ill patriarch.

When they saw his pale, shriveled face and his sunken eyes, Kell and Clanny immediately suspected the worst. Their fears grew as they heard the faint whispers of Sproat's story. Then Kell lifted one of Sproat's cold, pale hands. What had been the powerful hands of a boatman were limp and heavy. At the base of the thumb, Kell could feel the weak remnants of a pulse. As he knelt by the poor man's side, Kell's remaining doubts lifted and the horror of recognition took hold.

Kell had seen many strong, young British soldiers gripped by cholera. Just two years earlier he had been responsible for controlling an outbreak that struck a British regiment on the island outpost of Mauritius. Having seen cholera once, he could not forget it. Cholera had come to England.

More precisely, Asiatic cholera had arrived. The principles of British medicine that John Snow was dutifully learning just a few miles to the north painted a muddled picture of cholera. Hardcastle had taught his diligent pupil that good health required a proper balance of the four bodily humors: blood, phlegm, black bile, and yellow bile. The last of these, yellow bile, was also known as choler. Imbalance meant disease. Cholera, as noted in a medical book of the day, was "occasioned by a putrid acrimony of the bile."

Because of this fundamental misunderstanding, cholera's name was pasted across what we know today to be many different diseases, most of which involved severe vomiting and diarrhea. Snow had almost certainly seen cases of so-called common English cholera, a relatively nonspecific term for gastrointestinal diseases thought to be endemic to England. But as the epidemic approached from Calcutta, he quickly learned that this new sort of cholera was something entirely different. This new disease, according to the predominant medical thinking of the time, was a particular form of cholera known as either Asiatic cholera, to reflect its source, or cholera morbus, to reflect its severity. Today this is the only disease that we still refer to as cholera.

Most waterborne diseases cause diarrhea, vomiting, or both. They kill, ironically, by dehydrating their victims. Remarkably William Sproat hung on for several more days as cholera* sapped the life from his body. By Monday his blood began to grow thick and tarlike and his heart strained to pump the viscous fluid. Blood's most important role is to feed the fires of metabolism and haul away the smoke and ashes. As Sproat's blood slowed, those oxygen-starved fires dimmed. Slowly the color faded from his skin.

Kell had judiciously turned the care of Sproat over to Drs. Holmes and Clanny who did what they could, but medicine of the day had little to offer. The prevailing theories held that his body was trying to purge itself of some mysterious epidemic poison. Vomiting and diarrhea were to be encouraged. Recommended therapies routinely included emetics and enemas.

Three days later there was so little blood flowing to William Sproat's brain that he fell into a coma. As death moved in, his fingers and legs turned dark blue. That night Sproat died the "blue death" of cholera.

For the Sproat family, the tragedy was not over. Within hours, the

---

* For the remainder of the book, the term *cholera* will refer to the modern definition of the disease unless otherwise specified.

disease had the dead boatman's granddaughter in its grip. By the next morning it had reached out to take the poor girl's father.

As the week wore on, the number of cases rose steadily. The medical men of Sunderland held a meeting and concluded that cholera had undoubtedly arrived. They sent word on to London. In early November, health officers dispatched by the Privy Council in London had placed a quarantine of fifteen days on all ships originating *from* Sunderland. The epidemic was official and it had cut off Sunderland from the world.

But the fortunes of Sunderland, the fourth busiest port in England, rose and fell with the ships that filled her narrow harbor. A few weeks of unfavorable weather in the North Sea would send the local economy into a tailspin. An open-ended quarantine would cripple it.

Faced with this bleak reality, the medical men of Sunderland held a second meeting a week after the quarantine was announced. Kell and Clanny were not included. The assembled doctors, surgeons, and apothecaries declared that their earlier inference had been a misguided rush to judgment. They expressed their unanimous conclusion that the disease that had felled the Sproats (and several others in the days since) was not Asiatic cholera after all, but simply common English cholera. There was no epidemic.

The sloppiness of nineteenth-century diagnosis made this possible. Cholera, in 1831, was simply a set of symptoms. Today cholera is a specific disease and a definitive diagnosis is based exclusively on finding evidence that the *Vibrio cholerae* bacteria is responsible. In 1831 these doctors had no idea that such an agent existed. Microscopes were not a part of medical training and the notion that something undetectable to human senses might have the power to kill seemed ridiculous.

The ignorance of medical science made it possible to pretend that this was not Asiatic cholera. Overwhelming pressure from commercial interests made this sleight of hand expedient. Unfortunately renaming the disease failed to stop it. Over the next few days, the number of cases mounted and denial became untenable.

Just thirteen miles to the north, in the busy port of Newcastle-on-Tyne, an eighteen-year-old surgeon's apprentice attended to his duties. For four years Snow had ground and mixed medicines for Dr. Hardcastle, taken his messages, managed his appointments, and written up the daily entries in Latin. He had assisted Hardcastle in everything from pulling teeth to delivering babies. As he gained experience, he had begun to see some of Hardcastle's indigent patients on his own.

Snow followed the approach of cholera intently. He picked up information from any source available and tried to comprehend the convoluted web of primitive thinking as to its cause. Events in Sunderland also taught him about the capacity of the medical establishment to turn from the truth when economic and political forces make it desirable and ignorance makes it plausible, a process recapitulated in his own career many times over.

As November wore on, the medical community of Newcastle braced itself for the epidemic and John Snow attended to his duties. England was at war with cholera. Within a month, he would join the battle.

Cholera moved steadily north along muddy streets lined with human and animal waste. By late November it had reached the working class town of Tynemouth, just a mile downstream from Newcastle. It spread easily in this unsanitary world. The path upstream was short. On December 7, 1831, cholera arrived in Newcastle.

An unwashed hand, a dirty spoon, a bit of soiled linen. All innocent and harmless under normal circumstances. In the presence of cholera, however, they become the carriages in which death can ride. And so, in Newcastle, it did. Steadily from one person to the next. A touch, the sharing of a poisoned object, and the disease moved.

Epi-demos. Upon the people. An epidemic, by definition, afflicts large populations. Most epidemics, however, do not spread at a steady pace, but in fits and starts. Slow spread is punctuated by periodic, explosive outbreaks. In Newcastle cholera was picking off victims one by one. As Christmas approached, only a few dozen had died. Although the presence of the disease spread fear throughout the community, it had yet to show its potential for a sudden devastating outbreak. That was about to change.

As the residents of Gateshead sat to eat their Christmas dinners, Thomas Fife, a local apothecary, made his way through the narrow streets to a small, low-ceilinged room on Oakwellgate Lane. Like the other doctors of Gateshead, Fife had been following the news of the slowly growing outbreak in Newcastle, just across the river Tyne. As he walked through quiet streets of Christmas, Fife could hope that Gateshead, with only two deaths, might still be spared. A week earlier an impoverished ragpicker had died from cholera. The second case, a poor woman from Pipewellgate, had just fallen ill on Christmas Eve, but as Fife approached the tiny apartment of Margaret Taylor, he could still believe Gateshead would not feel the full wrath of cholera.

The air in the single room that Margaret Taylor shared with her sister, Isabella, was oppressive and still. The two women made a meager living as spinners at the twine yards. As he approached Margaret's bed, her sunken eyes gazed up from a face that appeared far older than her forty-two years. The illness had taken its warmth and vigor leaving it shriveled and gray. Margaret Taylor had been fighting for her life since four in the morning and was now struggling to breathe. Fife sat next to her and laid his fingers across the cold skin of her wrist. He waited in silent concentration, but her feeble pulse eluded his touch.

Fife saw that she had entered the later stages of cholera in which the disease seems to asphyxiate its victims. Concluding that he needed to stimulate her and to excite her vascular system, he gave her a combination of ammonia, camphor, opium, and menthol. As the cold blue death crept in, he tried to warm her by rubbing her with heated flannel and administering warm water enemas. Fearing he was facing his first case of Asiatic cholera, Fife sent word to his colleagues, including Tom Brady, the man who had cared for the first two cases in Gateshead. Soon several doctors including Brady crowded into the room. Brady confirmed Fife's conclusion that Margaret was in the final stages of the disease.

As Margaret Taylor lay dying, Fife was summoned to the bedside of a twelve-year-old girl, Mary Wheatley, who had been suddenly stricken with violent diarrhea and vomiting. Fife began to suspect that the epidemic had arrived in Gateshead as he rushed to examine the poor child. Relying on the confused concepts of disease and health that defined medicine in his day, Fife chose to purge her of the offending bile. He gave her ipecac to improve the effectiveness of her vomiting and an enema with turpentine in gruel.

Across town an apothecary by the name of Greenhow sat by the bed of Margaret Walker, the mother of nine children, who had been sick for three hours. When he had finished, he prepared several slices of toast, soaked them in vinegar, and coated them with black pepper. He then arranged them on her abdomen in hopes of improving her condition.

Throughout Gateshead, the story repeated itself over the course of Christmas Day as doctors, surgeons, and apothecaries rushed to offer useless remedies. With no real understanding of the disease, doctors selected their mode of treatment based on an amalgam of archaic beliefs interpreted according to the idiosyncrasies of their own particular misconceptions. As a result it is often impossible to discern the method in their madness. In treating the first two cases of cholera, Mr. Brady had spread a poultice of bran between two sheets and wrapped it around the legs of the victims. Another local doctor advocated the use of tobacco enemas during the course of the outbreak.

Fife's efforts to help his two patients eliminate cholera's poisons only hastened their death by dehydration. Margaret Taylor would not survive the night. Twelve-year-old Mary Wheatley died early the next morning. Greenhow's warm toast was laudable only in its lack of effect.

It did no harm, but his patient fared no better than Fife's. That same morning, the nine Walker children would lose their mother.

Fife did have one success. Just a few blocks from where the spinster, Margaret Taylor, lay dying, Joseph Laws, a twenty-four-year-old laborer, began to feel uneasy. Unsure of what to do to calm his queasy stomach, Laws sat down to a plate of cold mutton. He would soon discover that the stirring in his belly was not hunger, but cholera. When he called for Mr. Fife, the surgeon apothecary suggested he drink as much thin gruel as he wanted. By providing him with a mechanism to replace the fluids he was losing, Fife almost certainly saved Joseph Laws's life. He failed, however, to recognize that he had stumbled on the only effective treatment for cholera.

Fife was too busy to think much about the mechanisms of cholera's attack. At three in the morning of December 26, he was summoned back to Oakwellgate Lane to see Isabella Taylor, the sister of the spinster who had died the previous night. He found Isabella in a neighbor's apartment. Death haunted the air. The cholera-ravaged corpse of her sister, Margaret, lay next door in the room where she had died. Now the disease had moved to Isabella. She would die by nightfall. Before the next day was over, cholera had attacked 119 people and 52 of them had fallen to the blue death.

The frail, the old, and the very young were most likely to die at the hands of cholera. A desperate mother with bowls of chicken broth might naively save her child's life by staving off dehydration. In 1829 survival might have involved nothing more than avoiding medical care.

Like the other doctors in Newcastle, John Snow interpreted the outbreak in the context of the prevailing thinking about epidemic disease. The medical science of the day held that the spread of an epidemic disease such as cholera must involve a miasma, a foul-smelling airborne poison. An explosive outbreak such as occurred in Gateshead signaled the sudden presence of a particularly dense miasma in the afflicted community. Only the air could spread a disease this widely and rapidly. Snow would spend the rest of his career proving these theories wrong.

Although no one realized it at the time, Gateshead had just witnessed the tremendous capacity of water to spread cholera. In the months before the arrival of cholera, a severe water shortage had forced the local water company to pump water directly from the Tyne in order to ensure an adequate supply. Residents of Newcastle and the surrounding towns consumed bottles carefully filled with untreated river water. As the epidemic grew, the chance that cholera would crawl through the pumps and into those bottles increased. In the days before Christmas 1831, in the town of Gateshead, across the river from Newcastle, cholera had gotten into the water.

Cholera had established itself, but it was winter. In the cold, the disease slowed, hunkered down, and waited for better weather. With the arrival of spring, it began to travel again. As the summer warmed the coal-stained towns of northern England, the disease exploded all over the kingdom. The toll in Newcastle rose steadily.

In August 1832, with cholera raging, William Hardcastle called John Snow aside. Cholera, he told him, had spread to Killingworth, a coal town two miles to the north. In addition to his responsibilities in Newcastle, Hardcastle was Killingworth's only physician. At the same time, the vestry of St. John in Newcastle had recently given Hardcastle responsibility for overseeing the care of the poor in the parish. Given the new assignment together with his existing clinical responsibilities in Newcastle, he simply could not take on this new outbreak. After five years of training, Hardcastle believed Snow was ready to work on his own. At the age of nineteen John Snow set out for Killingworth to take on cholera single-handed.

The Killingworth Colliery was one of dozens of shafts sunk into the vast coal seam that ran under the northeast corner of England. Snow knew that cholera loved nothing more than a mining town. In January the initial wave of the epidemic had struck at Newburn, a mining village of 550 people. Within a few weeks, cholera had attacked 330 of them, killing 65.

Rotting garbage, excrement, and poor ventilation were all presumed to contribute to miasmas and the spread of epidemics. Had miasmas existed, they would have found a warm welcome at the bottom of a mineshaft. The miners worked and ate in filth that was unparalleled, even in the gritty world of the working class in early nineteenth-century England. There were absolutely no regulations to protect the men, women, and children that worked in the mines. (It would be eleven years before Parliament passed a law prohibiting women, girls, and boys under ten from working in the mines.) Miners routinely worked shifts of twelve hours or more. Whatever food and drink they needed for those twelve hours, they would take with them into the mine. There, in a dark narrow cave with no privy or water, they would live, eat, and defecate.

For John Snow the weeks he spent in Killingworth left the most enduring memories of his entire apprenticeship as he watched the miners emerge from the coal pits stricken with the disease. Snow worked tirelessly throughout the outbreak. Day after day he visited the one-room houses that stood in rows around the mine. Night after night he cared for the sick and dying miners and their families.

Despite his best efforts, he watched again and again as cholera wreaked sudden and deadly havoc on its victims. Away from Hardcastle's supervision, he had greater freedom to think for himself and to learn from his failures. At some point he began to sense that there was something fundamentally wrong in the prevailing wisdom about cholera.

Later, as he reflected on the outbreak, he wondered why cholera had struck at Killingworth while sparing other mining towns. With Robert Stephenson, the famed designer of steam locomotives, as its engineer, this mine was at least as well designed and ventilated as other mines in the area. What sort of miasma would strike at this mine and not others?

In the humid tropics, cholera becomes endemic, simmering in the population, limited only by the immunity of those who have survived the disease and the caution of those never afflicted. In colder climates, a bitter winter can pull the disease up by its roots and in 1832 that is precisely what happened. After killing more than twenty thousand Britons, cholera left England, but the question would stay with John Snow. For the next sixteen years, in cholera's absence, the question lingered as he moved on to other projects. He completed his apprenticeship with Hardcastle and then two more apprenticeships with surgeons in nearby towns. In 1836, after he had finished his final apprenticeship at the age of twenty-two, he walked home to York.

He had been gone for over seven years and much had changed in his long absence. The hard work of his parents had continued to pay off for his family. They now owned a farm on the outskirts of York and his older siblings were already on track to become teachers, clergymen, and businessmen. Two more Snows had been born in his absence. One of them, a brother, had died from an infection before John ever had a chance to see him.

The reunion did not last long. John, as always, was looking forward. The 120-mile walk from Newcastle to the family's farm was merely a warm-up. After visiting York, John Snow planned to continue on to London where he would enroll in medical school.

Even if he had taken the most direct route, the trip to London would have required him to walk almost two hundred miles, heading southeast along the Great North Road. Young, full of energy, and eager for the challenge, John bid good-bye to his parents and his seven siblings on a September morning in 1836. He began to walk, not toward London and the rising sun, but toward Liverpool, a hundred miles to the west. This was to be a grand adventure.

Snow had planned a circuitous route to London that would not only expand his universe, but would also take him to see one of the most important people in his life, his uncle and mentor, Charles Empson. Charming, dapper, and worldly, Empson had come to Newcastle shortly after the arrival of his nephew from York. Through most of Snow's apprenticeship, Empson owned a shop in Newcastle where he dealt in fine art, antiquities, and exotic shells and minerals, many of which he had collected during travels in South America. The shop was just a few blocks from Hardcastle's surgery. It was a hub for the cultural elite of Newcastle and Snow was a frequent visitor.

But the shop in Newcastle had closed its doors long before John Snow finished his apprenticeship. Empson left under a cloud. A note in a local paper stated that he had been "the victim of a cruel, malicious, and slanderous report, fabricated and propagated by a vile wretch whom he had employed as a tradesman." The precise nature of those accusations was never recorded, but they were sufficient to cause Empson to close his shop and move to Bath.

So after a stop in Liverpool, Snow turned south, toward Bath. Again he shunned the direct route, embarking on a tour through the Welsh countryside that added another hundred miles to his trip. He walked through beautiful mountains and past coal towns at least as dirty as those around Newcastle, coal towns that had been largely spared by the cholera epidemic. After walking almost four hundred miles, he arrived in Bath, found his way to an apartment above a small gallery at 9 Cleveland Place, and knocked.

When Charles Empson opened the door, John Snow had not seen him for two years. The two men spent their days together in Bath catching up on the events since their time together in Newcastle and discussing Snow's future in London. Given the expenses that lay ahead for Snow, it is likely that he received financial support from his increasingly prosperous uncle. Then, after a warm farewell, John Snow began the final leg of his journey.

For ten days Snow walked past the fertile fields of southern

England. The farms buzzed with the activity of the harvest. Carts laden with produce filled the roads. As he approached London, the traffic thickened. Coaches and carriages joined the farmers' carts. Finally the countryside gave way to Kensington Gardens and Hyde Park. One can only imagine how his young heart soared as John Snow walked past Buckingham Palace and into the center of the city that, perhaps more than any other, defined the nineteenth century. His trip ended on Great Windmill Street in Soho where he would continue his medical education as a student at the Hunterian School of Medicine. To get there he had walked almost five hundred miles.

As cholera faded from the consciousness of London, John Snow threw himself into his medical training. He did not forget cholera, although it would not return to England for twelve years. The disease would find him again, but it would also find that those years had made him a far more formidable adversary.

2

SNOW ON CHOLERA

John Snow was a mere teenager the first time he watched a patient's flesh turn pale and cold from the tourniquet's grip. Almost twenty years later, the lingering images still haunted his memory. Above all, he remembered patients in pain. He could still hear their screams as the razor-sharp blade sliced deep into the limb, shredding the protective haze of morphine. Each cut nerve brought a new wave of agony, but, as the incision revealed the bone, a new kind of anguish began. As his mentor, William Hardcastle, cut into the clusters of densely packed nerve endings that surrounded the bone, they translated the intrusion into a thousand rivers of anguish that overwhelmed the patient's very soul. The lingering passage through the dense mineral matrix seemed more like torture than surgery. Faces crushed by pain still howled across the years for relief, but that relief never began before the thud of the limb in the catch basket.

Surgical pain was Snow's grim tutor throughout his medical training. Since coming to London, he had spent endless hours in the operating rooms of Westminster Hospital. After finishing medical school, he had served as one of the hospital's house surgeons (the equivalent of today's surgery resident). He had participated in hundreds of surgical procedures and each one was a race against pain. A good surgeon was a fast surgeon. Every surgeon in Napoleon's army had to prove he could amputate a leg in less than three minutes. Until the winter of 1846, speed was the patient's only friend.

That year, on the morning of December 28, John Snow and two colleagues walked through the bitter cold that had gripped the city since Christmas Eve. The implications of what they had just seen consumed his thoughts. The fate of a rotten tooth had just revealed to them the future of surgery.

The tooth belonged to a female patient of Dr. James Robinson, a friend of John Snow. On that late December morning, Snow and his companions had observed as Robinson picked up his pliers, grasped the tooth, rocked it back and forth, and pulled. The astonished audience watched and listened in awe as the bloody tooth emerged. Their amazement arose not from what they saw or heard, but from what they did not. The patient had not cried out, moaned, or even flinched. Robinson's patient had no special resistance to pain. The incredible quiet had come from a bottle of ether.

Just nine days earlier, Robinson had given the first demonstration of ether in England. Until that moment, pain and the struggle to control it had pervaded the practice of the British surgeon. By the time Snow saw him in action, Robinson had performed half a dozen procedures and was, by default, the country's leading expert on the use of ether, a technique he had imported from America just weeks after the publication of the first paper describing its use appeared in Boston. On that late December morning, the three men stayed to question Robinson, drawing from his relatively shallow well of experience as the patient recovered.

They emerged from Robinson's office into the oppressive air of winter. Tens of thousands of stoves and furnaces that had burned through the cold heart of the British night filled the air with smoke. The coal-fired engines of London's factories added thick black clouds to the pall of soot that sank steadily downward on the populace. At the same time, another cloud was rising from the streets themselves. The steady hammering of hundreds of thousands of horseshoes was grinding the cobblestones into a fine powder. The result was a gritty mixture of stone dust and burned coal that accumulated in every corner of the lives and lungs of the city's inhabitants.

After ten years Snow had learned to live with London's abuses. As the three men parted company and Snow walked toward his home in Soho, he had far more on his mind than the city's cold, gritty air. On that morning in December, his thoughts were fixed on the miracle of silent surgery.

Snow knew that ether would change everything. He also recognized the danger inherent in ether's magic. Physicians would need to walk a fine line between the reemergence of pain because of too little ether and the disaster of an overdose.

The power of ether had impressed him, but Robinson's technique had been crude and his methods imprecise. The dentist had simply allowed the patient to inhale ether from a flask until she faded from consciousness. Snow's knowledge of chemistry told him that the evaporating ether cooled the inhaler. As it cooled, the amount of ether vapor rising from the flask and entering the patient's lungs decreased. This meant that Robinson had no way to know precisely how much ether he was giving the patient. Furthermore he had no real way to control the amount she inhaled during the operation. At a time when no one else in London seemed to see these problems, Snow had already begun to imagine how to solve them.

In the years since he arrived in London, Snow had turned his apartment into a buzzing laboratory, an outlet for his inquisitive mind. Even as a medical student, he had pieced together enough equipment to perform a study demonstrating that the arsenic used to preserve cadavers was rising from the dissecting table and sickening his fellow medical students.

At the time Londoners often used candles made from palm oil. Manufacturers added arsenic to help them burn brighter and more evenly. Snow's attempt to study arsenic levels released by burning candles almost ended his research career. He and a fellow medical student were hunched over their experiment, when an apparatus he had designed to collect the fumes from the candle burst into flames. Fortunately the two were quick enough to extinguish the blaze before it consumed Snow's nascent laboratory.

In the years since, his experimental technique had improved and his laboratory had grown to a point where he was ideally prepared to investigate the use of ether. He had the instruments needed to measure the temperature and pressure of gases with a high degree of precision. He had the cages and tanks he needed to test ether on a broad range of animals and, during the month that followed, he filled them with finches, thrushes, frogs, mice, rats, and even fish. He knew and had worked with high-precision machinists capable of turning his designs for medical devices into working prototypes.

So for the next month, he brought all his skills, experience, and laboratory resources to bear as he focused on the single goal of under-standing and refining the administration of ether. He devoted all his free time to the task and before two weeks were over he had the data necessary to predict the dose of ether over a broad range of conditions. Before the end of January, he had administered ether in precise doses to the menagerie of small animals in his lab and recorded their response in fine detail. At the same time, he had designed and assembled an ether inhaler intended to maintain a constant temperature and deliver a stable dose throughout a surgical procedure.

So by early February Snow had developed a detailed understand-ing of the effects of the drug, generated data essential for its controlled administration, and invented an inhaler far superior to any other. He arranged to present his findings at the regular evening meeting of the Westminster Medical Society.

It is no exaggeration to say that the society was the closest thing John Snow had to a family during his time in London. For much of his time there, he had no relatives in the city. There is no record of any romantic relationship during those years. His life was medicine. His friends were members of the society and he rarely missed their weekly meetings. From his first paper on the resuscitation of newborn infants to his last, he tested every new idea on his medical family.

Snow had presented his research to the meetings of the society before, but never with such overwhelming authority. Less than two months after first learning of ether's power, he had established himself as an expert in its use. His presentation impressed his colleagues and their acclaim came as a great triumph to the young doctor.

Snow however does not appear to have been motivated by any need for praise or money. He made no effort to patent his inventions. Instead he published detailed descriptions to allow others to copy them for their own use. Initially he did not even have notions of establishing a practice in the administration of ether. That changed on a January morning in 1847 as Snow was leaving the hospital.

Years later, a friend recounted the story:

> [H]e met a druggist whom he knew bustling along with a large ether apparatus under his arm. "Good morning!" said Dr. Snow. "Good morning to you, doctor!" said the friend; "but don't detain me, I am giving ether here and there and everywhere, and am getting quite into an ether practice. Good morning, doctor!" Rather peculiar! said the doctor to himself; rather peculiar, certainly! for this man has not the remotest physiological idea. An "ether practice! If he can get an ether practice, perchance some scraps of the same thing might fall to a scientific unfortunate."

At the time, Snow was, indeed, a "scientific unfortunate." He had tried to establish a clinical practice for more than eight years and was still struggling to make ends meet. He had contracts with several sick clubs, a system akin to a crude HMO in which doctors received a fixed fee and in return provided whatever care was required by members of the clubs. He also worked at the dispensary of Charing Cross hospital. In both cases, he was providing care to the working poor with minimal compensation.

To some extent, the cause of his financial struggle lay in the peculiar nature of the man. John Snow was an acquired taste. His voice, which a friend described as "painful" on first hearing, left a poor first impression. He also had the annoying habit of telling patients the truth rather than what they wanted to hear. In Newcastle his inclination to tell patients that Victorian medicine had little to offer them had resulted in bitter arguments with William Hardcastle and may have contributed to his decision to leave the city for further training. In London it left patients dissatisfied and cut into Snow's ability to generate an income. Last, but not least, he lacked the breeding, elite credentials, and connections necessary to attract the members of London's upper crust.

To make his financial matters worse, he had poured all that he could save of his meager income into research. That chance encounter with a druggist began a merger of Snow's research and his clinical work that would redefine his professional life. On January 28, exactly one month after observing Richardson's tooth extraction, John Snow put his first patient to sleep.

Over the next year, John Snow administered ether more than a hundred times for everything from mastectomies to dental procedures. His anesthesia practice opened a door to wealthier patients and brought a new source of income that allowed him to substantially increase his research activity. At a time before the specialty even had a name, he was on his way to becoming one of the most respected anesthesiologists in England.

But as his career developed, cholera was making its way across Europe to find him again.

After 1832 cholera had retreated to India where it festered until the hot summer of 1845 when it marched on Kabul. By 1846 it had reached into the Middle East, killing twelve thousand people in Teheran and thirty thousand in Baghdad. On a single, horrible night in 1847, almost three thousand pilgrims died at Mecca. In 1848 it made its way to a district on the banks of the Thames known as Horsleydown.

Named after a pasture used to graze horses in the sixteenth century, Horsleydown no longer appears on maps of London. Tower Bridge bisected the riverfront neighborhood when it was built in the late nineteenth century, relegating Horsleydown to the history books. In the 1840s however it was very much alive. An exotic blend of immigrants had settled there and established the breweries, tanneries, and warehouses that dotted the area. Above all else, its proximity to the docks made it the rough-and-tumble home to the seamen and dockworkers who serviced the endless stream of merchant ships that flowed into the city from all over Europe.

Two events have kept Horsleydown from disappearing entirely into the mists of history. The most dramatic was a fire that started in a jute warehouse on Tooley Street in 1861, near the wharf. A neighboring building containing resin, oil, and tallow exploded in flames and spewed its burning contents across the neighborhood. The resulting fire raged through the district for weeks, consuming the low wooden buildings in a blaze that could be seen for fifty miles.

The buildings that burned had provided the setting for a less dramatic, but ultimately more significant moment thirteen years earlier in September 1848. A seaman by the name of John Harnold had just returned from the port of Hamburg aboard the merchant ship *Elbe*. With cholera raging in the German port city, British troops placed the ship in quarantine.

Harnold must have been feeling a bit out of sorts, wanting nothing more than to lie down in a proper bed and rest. He could not tolerate the foul, cramped confines of a merchant ship when the city and freedom were so close at hand. In the busy, buzzing confusion of the docks, slipping ashore was far easier than the masters of quarantine liked to pretend.

So he made land and walked through Horsleydown to a rented

room on Gainsford Street, just two blocks from the docks. He lay down to rest, carrying the seeds of the epidemic in his belly.

When cholera struck at John Harnold, it did so with such force that the sturdy seaman died in just two hours. Within days the nightmare had begun to replay itself throughout the district. The second great epidemic of cholera had begun.

The disease that was once a stranger to London's physicians had taken on a terrifying familiarity. The first epidemic, which began in Sunderland in 1831, had killed 8,500 Londoners before ending in 1833. When Mr. Russell, an apothecary, came to see the stricken man, he knew in a moment what had happened. Dr. Parkes from the Board of Health confirmed that Asiatic cholera had arrived. The news raged through the city's medical community and quickly reached the busy anesthesiologist John Snow.

Fourteen years had passed since the young John Snow watched cholera slaughter the miners of Killingworth. In those years the puzzle of its cause had never lost its hold on him. During his years in medical school, first in Newcastle and then at the Hunterian School of Medicine on Great Windmill Street in Soho, he had learned the prevailing medical beliefs about cholera. Those theories however did not seem to fit the disease that he had come to know with a chilling intimacy.

The medical orthodoxy held that diseases were either epidemic or contagious, but never both. The epidemic diseases that had ravaged Europe since the birth of nations included the black plague, typhus, yellow fever, and malaria. Among the most terrifying features of these epidemics was their apparent ability to attack people who had no exposure to the disease. Contagious diseases moved slowly from one person to the next and could be contained by isolation. Epidemic diseases, on the other hand, could move with explosive speed. Quarantine could not contain them.

The dominant explanation for the cause of these epidemics held that they spread by a poison in the air, a miasma. Only an agent spread by the wind could move with such speed and defeat all efforts to stop it. Gaps in scientific knowledge with respect to the composition of the atmosphere allowed one to imagine the existence of an airborne poison. Malaria even took its name, which translates as "bad air," from this misconception. More than half a century would pass before scientists understood that the plague, typhus, and malaria rode in on rats, fleas, and mosquitoes, not bad air.

Cholera, with its demonstrated ability to devastate a wide area in short order, not to mention its apparent ability to slip through the most stringent quarantine, was, by definition, an epidemic disease. Classic medical wisdom held that, as an epidemic disease, cholera could not spread through contagion. In fact in the 1830s the *Lancet*, England's leading medical journal, published a series of articles that derided the foolish notion that cholera was contagious.

John Snow had a problem with this line of thinking. As a young man in Killingworth, he had watched cholera spread from miner to miner and then on to family members. He found nothing compelling in the arguments against contagion. However he had also seen the explosion of disease in Gateshead that occurred suddenly and simultaneously over a wide area. Cholera it seemed was both contagious and epidemic.

Cholera roamed the streets of London through the fall of 1848, killing at will. During those months, Snow's probing led him to a crack deep in the foundation of current medical thinking. As he burrowed deeper into that crack, the tired concrete of ancient beliefs began to crumble around him.

If an airborne poison spread cholera, then that toxin should behave in a manner similar to the toxic anesthetic agents that Snow had been testing with such rigor in his home laboratory. Every gas he had ever tested showed a consistent, predictable dose-response relationship. Every person or animal exposed to a given dose of anesthetic had roughly the same response. But cholera selected its victims from among families living in the same buildings with no apparent regard for the fact that they all breathed the same air. Even when cholera did strike, its severity varied. Half its victims managed to survive the attack, while some died in a matter of hours. Cholera's cause, whatever it was, acted unlike any gas he had ever seen.

Advocates of airborne causes also pointed to the pathology of the disease to justify their case. Cholera's victims, blue and breathless, appeared to have been asphyxiated. To many this seemed to provide clear evidence that cholera struck at the lungs. Snow looked not at how cholera ended, but how it began. In every case the disease attacked the gut first. What if cholera's agent was not inhaled, but ingested?

Here however Snow's thinking hit a wall. Cholera's choice of targets seemed to have even less to do with the food and drink its victims consumed than with the air they breathed. Some shunned mutton and others avoided vegetables. Some drank only brandy and others drank well water. If an ingested poison caused the disease, that poison had no consistent hiding place.

Then, as the epidemic stretched toward the winter, John Snow happened on a radical solution. What if a tiny organism, one too small to be seen with the naked eye, caused the disease? Such an organism could hide in any manner of food or drink. More importantly, if an invisible bit of stool found its way into the mouth of a family member or visitor, the process could begin again and the outbreak could continue.

In a world of stinking privies and foul cesspools, a world where the streets were crusted with manure, rotting garbage, and human excrement, a world filled with things vile, dangerous, and obvious, Snow's notion that something undetectable could wreak such carnage was among his most radical.

Snow was steeped in the science of his day and knew well that advances in microscopy had revealed an unseen world of strange creatures. However people seemed able to consume these mysterious "animalcules" with no effect. The medical establishment paid little attention to these insignificant, odorless curiosities. How could something so small have any consequence for humans?

Snow followed a radical line of thought to counter this objection. He proposed that most of these animalcules were indeed harmless, but some specific types of organisms or at least one type was far from benign. If that specific organism were capable of replicating in the human gut, then perhaps, when its numbers were large enough, it could cause the explosive diarrhea that characterized cholera.

But Snow's carefully constructed explanation had a hole. Ingestion seemed to require close contact with a previous victim. This theory left unexplained the explosive outbreaks that struck suddenly over broad areas. Then, in the fall of 1848, John Snow had an idea.

Few people in London paid as much attention to drinking water as did John Snow. That relationship began with a book. As an apprentice in Newcastle, he came across a copy of *Return to Nature or A Defence of the Vegetable Regimen* by John Frank Newton. This strange, rambling screed warns of "the dire effects on the human frame of animal food," drawing on everything from the writings of Cicero to the science of Justus von Liebig to justify a vegetarian diet. As a teenage apprentice, Snow must have found something compelling in the book, for he spent much of the remainder of his life as a vegetarian. This was not a casual undertaking in the unrefrigerated world of Victorian England. The long and literally fruitless winters were bland at best and threatened malnutrition at worst. Still Snow persisted at a time when this peculiar habit made him an object of curiosity.

Snow rarely did anything halfway and so he took on the book's other essential dietary advice. Not only was he an ardent and active teetotaler but from the time he first assembled a still in his quarters in Newcastle until the day he died, he drank distilled water. Newton's book asserted that even ordinary drinking water would undermine the benefits of a vegetarian regimen, primarily because of what he believed to be the routine presence of "animal oils" in drinking water. To Snow's early mentors, this was the strange and annoying habit of a confused adolescent, but Snow never relented. When he arrived in London, one of his first acts was to set up a still for his drinking water.

Snow's use of purified water may have saved his life as he found himself again and again at the epicenter of the cholera epidemic in England. It also offered a rare perspective on the water consumption of his countrymen. Perhaps it was during one of his distillation runs that the idea occurred to him. What if a drinking water supply were contaminated with fecal matter from a cholera victim? The deadly water could spread through a network of pipes and reach the water pitchers of unsuspecting victims over a broad area in a single moment.

His long string of cognitive leaps had landed on a solution. He did not have to decide if cholera was a contagious or an epidemic disease. It could be both. The dichotomy was a false one and now he had the mechanism to explain it.

John Snow's indictment of London's drinking water was not radical in and of itself. The medical community recognized that dirty water could make you sick. Snow, however, parted ways with his colleagues when it came to how water could make you sick. They held that fermenting organic matter in drinking water could release the same foul-smelling miasmatic poisons that caused epidemics and that these poisons could kill. Certainly, they conceded, drinking straight from the reeking water of the Thames could cause disease and might even spread cholera. Snow's theory required no such miasma. If his ideas were correct, water with no smell or visible contamination could kill.

John Snow had enough experience in the arena of ideas to know that a theory resting on so many untested suppositions would not carry much weight among his medical peers. He needed data. In the study of anesthesia, he had been able to conduct experiments to test his hypotheses. In his investigations of the cause of cholera, his superb skills in the laboratory could not give him the data he needed.

Animal experiments were not an option. Efforts to find the cause of cholera during the first epidemic had included feeding the excretions of cholera victims to a variety of animals. None of them had developed cholera. This was taken as proof that the cause of cholera was not ingested. Snow did not believe this, but had to find a way to disprove it.

Conducting human experiments was ethically impossible. To do so would have involved assembling two similar groups of people and exposing one group to contaminated water and one to pure water. If his theory was right, many of the subjects would have died in his quest for the truth. Snow could not conduct the trial, but he did have an alternative. He could wait for cholera to conduct the trial for him.

So Snow had to watch and wait. Cholera's experiment required a neighborhood with two water supplies. If it contaminated one and not the other, the resulting outbreak would point an accusatory finger at the contaminated drinking water. For Snow to begin to prove his theory, cholera needed to conduct the experiment. Unfortunately the deadly disease was not in the habit of publishing the results of its grim research. Instead, amidst the chaos of the epidemic, Snow would need to find the data.

Month after month, John Snow sat on his revolutionary idea. He broke his silence only once, when he tested the theory on two trusted colleagues. They listened with interest, but remained unconvinced by Snow's argument. Their skepticism did not discourage Snow, but it did convince him to wait until he could gather more data before presenting his theory to a wider audience.

For almost a year Snow's notion stayed within this small circle. During that year he developed a network of informants to help him track the disease's every move. As it struck one neighborhood after another, he visited the site of each cluster, talking to the doctors involved and to the survivors. Through the winter and spring, his investigations came up empty. Then in the summer of 1849 cholera showed its hand.

Cobbled together from pieces of London's history, the systems that carried water and waste had emerged from the city's agrarian past, evolving according to the expediency of the moment. There was no master plan and the architects of disaster rushed in to fill the void.

London's first residents had relied on the Thames and its tributaries for their water. As the city grew, many residents of London and the farms around it dug wells to find drinking water. Others built cisterns to capture rainwater. With further growth, water companies began to supply piped water, much of it drawn from the Thames. With the help of their friends in Parliament, the water companies carved up the city into districts over which they maintained exclusive control. By 1849 a Londoner might rely on wells, cisterns, piped water, or even buckets of river water. The choice depended on location, income, and personal preference. That choice would come to mean the difference between living and dying.

London's system for disposing of wastewater depended on the separation of waste. Feces and urine accumulated in the privies, which were placed at some distance from any residence. No amount of perfume or lime could make a trip to one of these Victorian outhouses palatable. One routinely emerged gasping for breath and praying for constipation. A nighttime trip was to be avoided at all costs, so a portion of the city's urine was stored under its beds in chamber pots during the night and flung out the window in the morning accompanied by a cry of "Gardez loo!," a bastardazation of the French *Gardez l'eau,"* or "watch out for the water." Wastewater from bathing, laundry, dishwashing, and household cleaning accumulated in the cesspools where water would seep into the ground and solids would slowly accumulate.

The guild of night soil rakers not only cleaned the privies or ash pits (euphemistically called because of the practice of spreading fire-place ash on them in an attempt to control odor), but also paid for the privilege as they could resell the valuable fertilizer to farmers on the outskirts of the city. Periodically they also cleaned the waste that accumulated in the bottom of London's cesspools. The rakers could never come often enough.

Sewers were intended to drain the streets. As the streets were far from clean, the sewers carried far more than rainwater. Grant's report on the outbreak at Albion Terrace concluded that the source of the problem was simply the miasma escaping from the rotting organic matter in the sewers.

The *Vibrio cholerae* bacterium is a creature that thrives in the warmth and moisture of a human intestine. Once out in the world, it struggles to survive long enough to find another victim. The warm waters of India suited the microscopic killer, helping it to find refuge between pandemics. For most of the year, the cold water of southern England was far less hospitable. Then as the summer sun warmed the Thames, London came to look, from the perspective of cholera, more and more like Calcutta.

The Surrey buildings, a cluster of small apartments in the dockside district of Horsleydown, drew their water from a well. During heavy rains, the well had a tendency to overflow. When it did the water would wash up onto the street before draining back down into the well. Whatever illness this caused before the summer of 1849 had drawn no notice. In late July of that year however, cholera was so widespread in this poor district on the Thames that two of the buildings' residents had the disease. Their family members had dumped the wash water from the sheets into the gutter. As the rain fell, it scoured the gutters and *Vibrio cholerae* poured into the well.

Eleven people in the Surrey buildings would die from cholera over the two weeks that followed. When Snow investigated and learned of the contamination of the well, he saw it as evidence of waterborne disease. But the residents of the Surrey building were poor; they were expected to die of cholera. The disease was easily found in the surrounding neighborhoods. Even though the rate of disease was far higher in the Surrey buildings, Snow suspected this evidence alone would not prove compelling.

At almost the same time however, cholera had infiltrated Albion Terrace, a far different neighborhood about a mile upstream and far removed from the squalor of the river's edge. When John Snow learned that cholera had killed twenty people from a row of seventeen upperclass homes in a matter of days, he rushed to the neighborhood to investigate. Albion Terrace was an island of devastation surrounded by a cholera-free sea.

During the ten days of the Albion Terrace outbreak, 1,231 people had died of cholera in London, but this outbreak among the affluent had already attracted the attention of the General Board of Health. When Snow arrived, their emissary, an engineer by the name of Thomas Grant, had already begun his investigation.

The residents had learned to live with flooded basements after each heavy rain. Mr. Grant's excavations had revealed how cesspools, swollen with rainwater, had overflowed and contaminated the row of interconnected cisterns that served the buildings. He pointed the finger of blame at the resulting stench together with the emanations from an open sewer some four hundred feet away. A miasma, Grant concluded, had caused the disaster.

When John Snow examined samples that Grant had provided him from two of the cisterns, he reached a far different conclusion. Even before he reached his laboratory, Snow sensed that he had the evidence he needed. As he picked through the foul sludge that Grant had scooped from a thick layer on the bottom of the tanks, he found the peel of a grape.

Something more than mere wash water had found its way into the drinking water. Unless someone in the building was in the habit of peeling grapes, the empty peel that Snow found had first passed through a human digestive tract. This meant that a bit of privy soil had found its way into the cistern and with it the seeds of cholera.

Snow rushed to put his theory to paper. Night after long night, he assembled his case. As he wrote, the undigested bits of food in the specimens from Albion Terrace that sat in his office fermented with such ferocity that the corks would pop out of their bottles on the shelves above him.

Less than a month after the outbreak, he walked into the office of John Churchill and Sons with a manuscript under his arm. By early September the Soho publisher had provided Snow with a stack of the thirty-one page monograph still smelling of fresh ink. Having conducted the research and paid for its publication, he set to work distributing copies of *On the Mode of Communication of Cholera* to colleagues and medical journals.

If Snow had hoped for a quick acceptance of his theory, the response to his monograph was disappointing. Even though the notion that cholera was contagious had grown far more acceptable since 1831, London's two leading medical journals seemed unimpressed by Snow's efforts. The review in the *Lancet* was only two paragraphs long, one of which was simply a quote from the monograph. In this cursory treatment, the editors asserted that Snow's arguments against an airborne cause were "not by any means decisive." Recognizing that the review was too short to fully present and rebut Snow's work, they suggested that their readers refer to the essay, but cautioned that it "must of course be received with great limitation."

The *London Medical Gazette*, London's other leading medical publication, published a far more extensive, but no less dismissive, review. The review discarded Snow's analysis of the Albion Terrace outbreak, stating:

There is, in our view, an entire failure of proof that the occurrence
of any one case could be clearly and unambiguously assigned to
the use of the water. . . . Foul effluvia from the state of the drains
[i.e., an airborne miasma from the sewers] afford a more satisfactory explanation of the diffusion of the disease.

It went on to close with a wisp of faint praise:

Notwithstanding our opinion that Dr. Snow has failed in proving
that cholera is communicated in the mode in which he supposes
it to be, he deserves the thanks of the profession for endeavouring
to solve the mystery.

Snow pored over the reviews, looking not for affirmation, but for
a hint as to what it would take to convince his audience. The *London
Medical Gazette* spelled it out:

The *experimentum crucis* would be, that the local water conveyed to
a distant locality, where cholera had been hitherto unknown, produced the disease in all who used it, while those who did not use
it escaped.

Even as he shot back at the *London Medical Gazette* with a letter cor-
recting the errors in their review, John Snow had begun to search for a
place where cholera and chance had performed an *experimentum crucis*.

Snow pored over any and every publication that described the
impact of cholera on communities throughout the kingdom. He read
through the voluminous reports of the registrar general that tabulated
the deaths from cholera throughout the epidemic, searching for areas
where cholera had either taken an extreme toll or hardly visited. In
each case he paid particular attention to the water supply, searching for
cholera's path. Where the reports lacked detail, he sent letters to local
physicians seeking their help.

Example after example poured into Snow's small apartment on Frith Street in Soho. One report came from his hometown of York where a terrible outbreak had struck the narrow lanes along the River Ouse where he had spent his childhood. He could still remember watching neighbors draw their water from the river. The deaths from cholera in York, he learned, came to an abrupt halt when local health officials brought in water from far upstream and began anew when the importation of water stopped.

Snow made note of other towns where a change in water supply had accompanied a change in the death toll. Exeter, which had seen 345 cholera deaths in 1832 during the first epidemic, saw only 20 in 1849. Between the two outbreaks the town, which had been using a polluted millstream, built a new waterworks in cleaner waters far upstream. Hull, on the other hand, had seen a sixfold increase in cases after moving the water supply from small streams in the hills to the river flowing through the center of town.

As new reports came in, Snow assembled them into a paper that would give far more evidence to support his theory. Less than a month after the *Lancet* and the *London Medical Gazette* had printed dismissive reviews of his original monograph, he had presented his new findings to the Westminster Medical Society. At the time medical journals often carried the proceedings of medical society meetings. Two weeks later the *London Medical Gazette* published the first half of a two-part report by John Snow, "On the Propagation and Mode of Communication of Cholera."

By the time the second half of the report appeared a month later, on November 23, cholera had lost its grip on the city. After a high of more than 1,532 deaths during the week of August 7, cholera mortality declined steadily through the fall. During the last week of November, only one Londoner died of cholera. It was to be the final death of the epidemic.

Even after its disappearance, John Snow continued his work on cholera, but at a less frenzied pace. Cholera had disappeared for fifteen years after the last epidemic. To all parties it seemed there would be plenty of time for a full examination of the evidence and a reasoned debate on the prevailing theories.

Everyone had something to say on the subject. In just one issue, the *London Medical Gazette* carried reviews of seventeen different monographs on the causes and treatment of cholera, together with the first half of Snow's article, and four other articles on the disease. Only Snow argued that drinking water played a key role in spreading the disease.

Snow hoped that the many official reports on the epidemic, all written with far greater resources than he could muster, would unearth additional evidence to help prove his theory. But he underestimated the ability of those threatened by the truth to weave armor from twisted facts and distorted logic.

Snow understood that he was laying siege to the entrenched beliefs of the medical establishment, but their opposition paled in comparison to the financial and political forces marshaled against him. If Snow were proven correct about the ability of drinking water to transmit disease, the anticompetitive cartel that allowed London's water companies to enrich their stockholders with little attention to water quality might crumble. If his belief that cholera was contagious were true, quarantines would have their ultimate justification, a conclusion that put him squarely at odds with the vast economic interests that relied on international trade.   Ironically the greatest source of opposition to Snow's ideas came from politicians bent on protecting the public health. Snow's ideas put him on a collision course with a group that came to be known as the sanitarians. At their head was a great bull of a man whom John Stuart Mill called the most effective politician of his time. His name was Edwin Chadwick, and his perspective before, during, and after England's second cholera epidemic in 1848–1849 was far different from Dr. Snow's.

3

ALL SMELL IS DISEASE

On a cool September morning in 1840, Edwin Chadwick descended the winding streets of Glasgow into one of the city's most notorious slums. As he and his three colleagues snaked down from Argyll Street along the narrow alleys that lead toward the River Clyde, a world of decay closed in around them. More than one sixth of the cases in the fever hospitals of Glasgow during the past year had come from this single district. A local expert, Dr. Robert Cowan, was leading the four men on a search for the causes of this pestilence. As they approached the river, he turned and ducked through a low doorway. Chadwick followed, squeezing his burly frame through the opening, unprepared for what he would find.

The passageway opened onto a courtyard unlike any he had seen before. In its center rose a vast dunghill, the product of the destitute who crowded into the rooms around it. Relying on a strong stomach and an unrelenting will, Chadwick made his way along a narrow path that skirted the courtyard and passed through a second corridor on its far side. He walked through it and found, to his horror, a second courtyard, identical to the first with a second immense pile of human feces.

As Chadwick and his well-dressed entourage continued their tour, the denizens of this grim world, clothed in rags and tatters, crowded together for warmth, looked out at the intruders through hungry eyes. Unafraid to look this broken world in the face, Chadwick directed the men into one of the crumbling buildings. The desperate conditions inside the buildings matched the horrifying state of their courtyards. In one room, they found a group of women, huddling naked beneath a blanket. Their clothes, the men learned, were in use by their roommates. Without enough clothes for all the occupants of the room, the women took turns using what they had to venture out in the cold. The four well-dressed men had soon seen enough and returned to the courtyard.

The narrow path around the second courtyard led to a third passageway on its far side. Again Chadwick and his colleagues entered the portal and braced themselves as they navigated the dank, dirty tunnel. Chadwick emerged to find himself in a third courtyard with another dunghill at its center. As he gazed at this monument to human desperation, he must have felt that he had slipped from reality into a world sprung from the mind of Dante.

In the course of their tour, the men paused to ask the inhabitants why they allowed their own waste to accumulate in such an appalling manner. It was then that Chadwick learned that the vast piles of human excrement in every courtyard were not simply the product of sloth. The residents of the buildings that surrounded them had created the three mounds and retained them for a purpose. When the piles grew large enough, they summoned the night soil rakers of Glasgow who carted them away for sale to local farms as fertilizer. For this pile of their own manure, they received a pittance from the rakers, an amount they relied on to pay their rent.

As secretary of the Poor Law Commission, Chadwick was preparing a report on the living conditions of the nation's working class. This netherworld was only one of the kingdom's many darkened corners that Chadwick visited in preparing his report, although he would recall it as the worst. Even before he began his tour, Chadwick was convinced that, as he put it, "All smell is disease." Given the horrors he experienced in Glasgow and elsewhere, and the many similar reports he received from correspondents throughout Great Britain, one can forgive his unshakeable adherence to this belief. His intransigence however was to have disastrous consequences.

A strident utilitarian, Chadwick saw the diseases that afflicted the working class as a vast drag on the efficiency of the nation. Mortality rates, which had declined throughout the eighteenth century and into the early nineteenth century, were now rising. Between 1831 and 1841, they had risen fifty percent. The increase was worst in the rapidly growing urban areas.

The report grew out of an effort to address this crisis and no one could match Chadwick when it came to assembling government reports. He had put himself through law school by working as a journalist and could spin tales with relative ease. Back in his London office, Chadwick immediately set to work. Work was a reliable old friend. He would spend most of his life working ten- to twelve-hour days, rarely taking a day off. To prepare this report, Chadwick worked even longer hours, compiling stories to support his case.

Chadwick filled the 457-page *Report on the Sanitary Conditions of the Labouring Population of Great Britain* with example after example of the horrifying conditions of the country's poor. He concluded that the diseases that afflicted them were propagated by "atmospheric impurities produced by decomposing animal and vegetable substances." Odor was the problem and, Chadwick proposed, water was the solution.

According to Chadwick the disorganized and ineffective system of carting away the vast accumulation of dung and rotting garbage that was choking the towns and cities of the kingdom had to be replaced. With a steady supply of water, the cities of the kingdom could literally wash away their troubles. By his accounting a water-based system would cost less than half as much as the army of carts it would replace.

The water closet played a key role in Chadwick's plan. This early version of the toilet had been available since its invention by Sir John Harrington in 1596. At the time the proud inventor had installed one in his house and one in the house of his godmother, the Queen of England. Even with this royal endorsement, it would spend almost three hundred years as little more than a curiosity. Harrington's invention, it seems, had a flaw. It relied on water but, unless you were the queen, you were unlikely to be able to get rid of the water. If Chadwick wanted to enlist Harrington's invention to help remove London's waste, he had to solve this problem.

In our world of modern plumbing and sanitary sewers, the toilet's magic is almost invisible as the water flows underground to far off sewage treatment plants. In the early nineteenth century however, landlocked cesspools could not tolerate a flushing toilet and any effort to connect them to a sewer could send the enterprising plumber to prison. Sewers were reserved for rainwater.

From the time Henry VIII passed the Bill of Sewers in 1535 until the early nineteenth century, sewers were intended to drain the streets. Putting sewage into them was unthinkable. The word *sewage* would not even exist until 1849. In 1842 London's rainwater flowed through the sewers while the city literally sat on top of its wastewater—buried beneath London were more than 300,000 decaying cesspools.

Chadwick seemed to have an answer for everything and this was no exception. Under his master plan, household wastewater and the water from flushing water closets would flow directly into a new system of earthen pipes that would replace London's existing sewers. Cesspools would be eliminated. Routine flushing of the new sewers with a steady supply of water would keep the neighborhoods free from "atmospheric impurities."

Chadwick saw the answer in the sewers, but he still had to find a place to send all that dirty water. His long-term solution involved a system to carry the nutrient rich wastewater to outlying farmland for use as fertilizer. In the short term however, London would send a growing stream of raw sewage into the river that flowed through its very heart, the Thames.

Chadwick understood that sending thousands of tons of rotting waste into the Thames was not a perfect solution, but perfection would need to wait. In the short term, polluting a river that was already contaminated by the slop that ran off the city streets was a minor inconvenience compared to achieving his sanitary goals. As he put it:

> The chief objection to the extension of this system is the pollution of the water of the river into which the sewers are discharged. Admitting the expediency of avoiding this pollution, it is nevertheless proved to be of almost inappreciable magnitude in comparison with the ill health occasioned by the constant retention of several hundred thousand accumulations of pollution in the most densely-peopled districts.

Chadwick's ringing, dismissive phrase, "inappreciable magnitude," haunts us to this day. Even those who held to the sanitarians' views about smell and disease would, in a few years, concede that Chadwick had miscalculated. The return of cholera would turn Chadwick's blunder into a public health disaster.

In the hands of a lesser politician, Chadwick's vision might have never left the printed page. The scale of the project was unprecedented and the political obstacles immense, but he was never one to shy away from a task, even if it required redefining the government of Metropolitan London.

The sprawling metropolis was an unruly patchwork of countless local fiefdoms. There were eight different sewer commissions in the metropolis with more than 240 members. The streets under which Chadwick's improvements would run were the province of more than 300 parishes, improvement commissions, and boards of trustees controlled by thousands of influential local politicians. Chadwick sought the power to eliminate them all.

A large, imposing man with patrician roots, an abrasive personality, and a grand swoosh over his balding head, Chadwick had risen to prominence as a bureaucrat on the force of his will and the strength of his convictions. He exhibited little reluctance to offend. If bringing his proposal to life required taking on the entrenched power structure, so be it. He barged forward, driven by the absolute conviction that he was acting in the public good. His colleagues on the Poor Law Commission, however, were not so fearless. They refused to have their names on the report, leaving Chadwick alone on the title page.

Chadwick turned that isolation to his advantage. He focused his powers of promotion on getting the report into as many hands as possible. Chadwick's efforts to market the report, together with its shocking revelations about the appalling living conditions of the working poor, sparked unprecedented demand. Its printing exceeded ten thousand copies, dwarfing anything the government had ever published. In a single stroke, the report defined the sanitary movement, imbued it with great political power, and placed Chadwick at its helm.

The report's broad distribution also helped bring a vast army of influential supporters into the ranks of the sanitarians. From Disraeli to Dickens, from Florence Nightingale to the queen's physicians, Londoners had little difficulty accepting the need to purge the city of its stench and rallied behind Chadwick. That support led, in December 1847, to the creation of a single Metropolitan Commission of the Sewers with Chadwick in control.

A plan conceived by Chadwick and his ally in the legislature, Lord Morpeth, also dealt with the problem of the existing sewer commissions. Rather than eliminate them, the two men had a plan that would simply make the six most critical sewer commissions irrelevant. First they packed the newly formed Metropolitan Commission of the Sewers with devoted sanitarians. At the same time they arranged for these men to receive simultaneous appointments to six existing local sewer boards. By doing so, Chadwick replaced more than six hundred local bureaucrats with a small, select group of his followers.

Chadwick immediately launched a crusade to eliminate London's cesspools. One of the Metropolitan Commission of the Sewers' first new regulations required direct connection to the sewers in any new construction. In the spring of 1848, work crews spread out across the city, tearing up streets to begin the construction of new sewer lines.

Never satisfied, Chadwick wanted more power to implement his grand vision. Control of the sewers was not enough if he was to create his sanitary city. He needed a way to ensure there was enough water to flush his new sewers. Riding a wave of popular support, Chadwick and his political allies convinced Parliament to create the General Board of Health in September 1848. With no scientific or medical training, Chadwick was one of three commissioners. The Board of Health would have as its mandate the removal of the local sources of miasma, a task that would require water.

Chadwick worked at a feverish pace to realize his vision. Experience had taught him that political ascendancy did not last. When cholera slithered off the docks of London to begin the second epidemic in the fall of 1848, he responded by pushing his agenda even harder. As a result between the spring of 1848 and the summer of 1849, the flow of sewage into the Thames doubled. Chadwick's "improvements" had some unintended consequences. The first cholera epidemic killed one of every twenty-seven Londoners in a two-year period. The second would kill one of sixteen.

John Snow and Edwin Chadwick never met. Chadwick viewed Victorian medicine with something approaching disdain, an opinion not entirely unjustified. His medical contacts were limited to a few physicians who shared his devotion to the sanitary movement and provided the necessary scientific cover. However, as the wastewater of

London poured into the Thames, the vision of Edwin Chadwick collided with the wisdom of John Snow in ways that changed the lives of both men.

John Snow looked with some horror on the agenda of the sanitarians. The same river that Chadwick saw as the solution to the problem of disease in London, Snow recognized as a primary source of the city's worst scourge. The Thames flowed not only through the center of the city, but through homes throughout London.

Nine different companies sent forty-four million gallons of water each day into the British capital. Almost ninety percent of homes had access to this supply. Some of the water came from relatively pure sources such as the springs and artesian wells that supplied the Hampstead waterworks, but these supplies were limited. Most of the water came from the rivers that flowed toward and through the city. Of the nine suppliers, five drew their water from the Thames.

Many limited their use of piped water to washing and cooking, relying on well water for drinking. Over time, however, the thought of trekking to the nearest well and lugging back buckets of drinking water had the remarkable ability to purify the water that flowed so effortlessly from the tap. Ease of use nurtured a dangerous delusion. With sewage flowing into the Thames, tap water turned deadly.

As a stranger to the strange land of politics, Snow refrained from criticizing Chadwick directly. In his original monograph, *On the Mode of Communication of Cholera*, his critique of the sanitarians was limited to the advice that people avoid drinking "water into which drains and sewers empty themselves." Snow preferred evidence to policy critique. The report demonstrated that the rate of cholera mortality among people living south of the Thames was seven times higher than the rate on the north side of the river. All the water supplied to the south side, he noted, came from the Thames.

In the fall of 1849, emboldened by his growing body of evidence and outraged by the actions of the Commission of the Sewers, Snow took more direct aim at Chadwick and his allies. The update of his monograph published in the *London Medical Gazette* described cholera's devastating assault on the residents of Rotherhithe who drew their water from tidal ditches filled by the Thames. In many cases the ditches received the contents of surrounding privies. Seeing the consequences of drinking what he called "Thames water rendered a little richer in manure," Snow could not hold his pen. What the victims had drawn from these filthy ditches was, he pointed out, "probably equal to what Thames water would be if certain of our sanitary advisers could succeed in washing the contents of all our cesspools into the river."

Through his outrage Snow also saw that the increased pollution of the Thames had set in motion an experiment that might give him the final piece to his puzzle. If he could show marked differences in cholera rates among customers of different water companies according to the source of their water, particularly when those customers lived side by side, his critics would lose their refuge.

The retreat of cholera following the epidemic of 1848–1849 allowed John Snow to focus his energies on his own career and success built on success. In the spring of 1850, John Snow sat before the senior members of the Royal College of Physicians and endured a barrage of questions on classic medical texts. Passing the exam would make him a licentiate of the Royal College. His examiners sat at the peak of a hierarchy so rooted in the past that, until that year, the entire exam was given in Latin.

British medicine no longer intimidated the unflappable Dr. Snow. When the day was done, the college nodded its approval, granting him the highest medical credential available to a man with his plebian pedigree. Becoming a licentiate was a great achievement, but Snow could not hope to climb higher. The peak of British medicine, Fellowship in the Royal College, was the exclusive province of graduates of Cambridge and Oxford.

Nonetheless his success gave him entrée to exclusive medical societies and put him on a path that would lead to the presidency of the Medical Society of London in 1855. John Snow's acknowledged skill as an anesthetist also meant a dramatic improvement in his finances. In 1852 he moved from his apartment on Frith Street in Soho to a home of his own at a prestigious address on Sackville Street, just off Piccadilly. His income even permitted him to hire away his former landlady's housekeeper.

But no measure of success could compare to a single call for his services late in the winter of 1853. A woman living across from St. James's Park had experienced a series of difficult deliveries and had heard that Dr. Snow's anesthesia could ease her pain. She and her husband wanted to meet with him to discuss their options. Her name was Victoria.

It is hard to say which Queen Victoria disliked most, pregnancy, childbirth, or infants. Most likely she viewed them together as a troika of misery that afflicted women with oppressive frequency. That the interminable wretchedness of pregnancy and the agony of labor produced such a "nasty object" as a baby seemed a cruel joke. She viewed infants as "mere little plants for the first six months" characterized only by that "terrible frog like action." Both pregnancy and the postpartum period were marked by depression mixed with fits of irrational rage directed at her beleaguered husband, Prince Albert.

Despite her feelings she spent more than half of the first ten years of her marriage in what she called the "unhappy condition" of pregnancy and almost seven of those years with infant children. Before the birth of her seventh child, Prince Arthur, in 1850, Victoria and Albert had asked that her obstetrician consider anesthesia during labor, a practice frowned upon by much of the medical establishment. At the time the queen's obstetrician, Dr. Charles Locock, had sought out and consulted with the nation's foremost authority on the administration of chloroform, John Snow.

The dramatic appearance of ether on the medical stage had sparked a frenzied search for other anesthetic agents. Snow had been among the most rigorous researchers in this effort, testing dozens of agents not only on his small zoo of laboratory animals, but also on himself. He had been among the first to recognize that chloroform, which produced an anesthesia less overwhelming than ether, had advantages in certain procedures, particularly childbirth.

The queen's physicians considered anesthesia too risky. At their urging she had forgone anesthesia for the birth of Arthur, but now she was pregnant again. She and her husband had had enough of labor. Taking matters into their own hands, Victoria and Albert decided it was time to meet Dr. Snow themselves.

Almost twenty years after he had first marveled at the sight of the royal residence as a young surgeon, Dr. John Snow walked through its ornate entrance to meet his queen. Although the details of the meeting are lost, he must have impressed the royal couple. When the time came, the queen overruled her physicians and insisted on anesthesia.

On the morning of April 7, 1853, John Snow received an urgent message from Sir James Clark requesting his presence. The queen was in labor. Snow rushed down Piccadilly and across the St. James's Park. When he arrived at the palace, he was ushered to a sitting room a few discreet steps from the queen's chambers where her doctors had gathered. After two hours Victoria called for relief.

With the utmost care, Dr. Snow measured out fifteen minims of chloroform, poured it onto a handkerchief, and held it to the queen's nose. As she lost her connection to pain, he monitored her every breath and heartbeat. Within an hour Prince Leopold was born.

Many in the medical community considered it scandalous that the queen should have chosen to interfere with the natural course of childbirth. The *Lancet* responded to the news with a scathing editorial. After pointing to the deaths associated with the administration of chloroform, the editors opined, "[W]e could not imagine that any one had incurred the awful responsibility of advising the administration of chloroform to her Majesty during a perfectly natural labor." The editors considered this use of anesthesia "unnecessary." The queen held no such opinion. Chloroform was, in her estimation, "soothing, quieting, and delightful beyond measure." Four years later, when her ninth and final child was born, she would make sure that Dr. Snow and his "blessed chloroform" were by her side.

Cholera had departed, but it left Londoners reeling from its impact and struggling to understand its causes. Early in the spring of 1852, a stout, balding man looked out with dark, sunken eyes over a crowd that had gathered around his podium just off St. James's Square. They listened intently as he spoke out on man's natural place in the broad sweeping uplands of the world:

> An instinctive sense draws [man] to the healthy places of the earth, and makes the lands in which his race dies and is degraded, repulsive. In dank marshes surrounded by stagnant waters, and in hollow places of the earth covered with reptiles, we feel oppressed; on the plain, where the breezes sweep over the herbage, the mind as well as the body is at ease.

Just prior to expounding on the healing virtues of the windswept highlands and warning against the race-degrading dangers of low places, the speaker had presented a precise mathematical formula to predict the rise in cholera rates as populations descend toward the sea. He next launched into a sweeping thesis on the deleterious effects of the lowlands throughout history. In the higher altitudes, he argued:

> [T]he chest expands in the elastic air, and the soul seems to drink in deeper draughts of Life. On the high lands men feel the loftiest emotions.... Man feels his immortality in the hills.

He explained how miasmas collect in lower altitudes and laid out the consequences of settling in such undesirable settings. "Wherever the human race, yielding to ignorance, indolence, or accident, is in such a situation as to be liable to lose its strength, courage, liberty, wisdom, lofty emotions," he warned, "the plague, the fever, or the cholera comes." These unhealthy environs lead not only to disease, but also, over the course of generations, to a profound degradation of the population. He presented examples from India to ancient Rome to demonstrate the tendency for races that had descended toward the oceans and estuaries of the world to enter a slow, steady decline.

Clearly the English people had come from the healthful highlands and belonged there still. "No race of men living in maremmas, marshes, deltas, low sea-coasts, low river-sides, could have acquired or wielded the power of this empire," he observed. Cholera was perhaps a clarion call for a return to pastoral roots, a bit of tough love from the deity. After all, he observed, "to a nation of good and noble men, Death is a less evil than Degradation of the Race."

The attentive crowd that stayed to the end of this diatribe on the preservation of the English race was not a fringe group, but the Statistical Society of London. The speaker was not an eccentric, but William Farr, the leading medical statistician of his day. As the Compiler of Abstracts for the General Registrar, he was the man responsible for the tabulation of all vital statistics, including mortality rates. No one knew the cholera statistics better than he. His report based on the principles outlined in this speech became one of the most widely cited works on the cause of cholera in his time.

Farr had been refining this idea since the epidemic of 1848–1849. He was committed to the sanitary cause and the sanitarians had relied on his observation to dismiss Snow's findings about the relationship between water supply and disease. Snow, they argued, had correctly described the occurrence of cholera, but had identified the wrong cause. The districts that relied on the Thames for their water supply were lower in altitude. Where Snow saw the effects of contaminated drinking water, Farr, along with Chadwick and his allies, saw miasmas sinking toward sea level.

For the sanitarians the devastation of the epidemic of 1848–1849 demonstrated not that they had been mistaken in dumping the city's sewage into the river, but that they had not moved fast enough in purging the city of fermenting organic matter. Through the early 1850s, they redoubled their efforts. By 1855 thirty thousand cesspools had been eliminated and far more had been bypassed. The amount of sewage pouring into the river soared as Chadwick pressed his agenda.

Still the truth has an annoying resilience. The sanitarians did not need to accept Snow's thinking, but they could not entirely ignore it. In 1850 the General Board of Health issued an investigation of the city's water supply. The report made no direct mention of Snow, but implied that the sanitarians had always recognized the link between water quality and health. Of course, it explained, excessive organic matter in the water supply will ferment and contribute to the miasma. Even Hippocrates had warned of the ill effects associated with certain types of water. Snow had simply taken an obvious relationship and carried it too far with his strange notions about ingestible, self-replicating agents of disease.

The report also argued for the consolidation of the nine private water companies into a single entity, controlled by the government with a single source, far from the tidal reach of the Thames. In extolling the advantages of the supply they had identified, the board spent more time on the superior softness of the water than on any sort of biological purity. By avoiding the mineral content of the Thames, they boasted, the new water system would save five million pounds per annum in soap and cut tea consumption for the metropolis in half.

Once again Chadwick's report created an alliance of powerful enemies. The water companies had established a formidable oligopoly through a series of anticompetitive agreements. This lucrative arrangement had filled their coffers and given them the guaranteed allegiance of at least a hundred members of Parliament. By the time the Metropolitan Water Act passed in 1852, Sir William Clay, spokesman for the water companies, had been enlisted to draft the final bill. It gave the water suppliers until 1855 to cover their reservoirs, filter their water, and move any intakes in the Thames upstream, to put them out of reach of the tides. In return the water companies were left standing. Chadwick however, with too many enemies, stood on the brink of political destruction.

When John Snow held his chloroform-soaked handkerchief to Victoria's nose in the spring of 1853, both he and England were at the height of their powers. With her rich coal reserves and vast merchant and military fleets, Britain dominated the industrial revolution and the international trade it generated. Just two years earlier, the world had made a pilgrimage to the Crystal Palace, a spectacle of glass and steel with a monumental glass fountain and three fully grown elm trees at its heart. Erected in the center of Hyde Park, its two vast arms sheltered the Great Exposition. The two arms stretched out in perfect symmetry, balanced like an immense, glittering scale. In one arm it held the wonders of the British Commonwealth. In the other the accomplishments of the remainder of the world made the perfect counterweight.

The nightmare of the recent cholera epidemic was fading into history. It must have seemed that the collective expertise of the magnificent empire could solve the riddle of its cause and prevent a third epidemic. Any such illusion would soon be shattered.

Snow's prestige as an anesthesiologist did not translate into acceptance of his ideas about cholera. In 1853, the same year he attended to the queen, the editors of the *Lancet* expressed their frustration at the lack of a satisfactory explanation for cholera's cause:

What is cholera? Is it a fungus, an insect, a miasma, an electrical disturbance, a deficiency of ozone, a morbid off-scouring of the intestinal canal? We know nothing; we are at sea in a whirlpool of conjecture.

As an epidemiologist, Snow had to rely on the epidemic for his data. Cholera's retreat in 1849 had halted its natural experiment, but Snow kept working. For five years, as he maintained a burgeoning anesthesia practice and a busy agenda for research on new anesthetic agents, he assembled data. He learned everything he could about London's water suppliers: where did they draw their water, how did they filter it, if at all, and whom did they serve? His analysis of these data would be the centerpiece of a revised monograph. He had nearly completed his study of the first two epidemics when in 1853 cholera began to experiment again.

4

THE *EXPERIMENTUM*
CRUCIS

As the steel gate slammed shut behind him, a lone figure emerged from the gloomy corridors of Millbank Prison and crossed the bare, graveled wedge of the outer yard. At its far end, he entered a triangular room occupied by a single table. On it was a list of those who would be leaving the prison that day. As William Baly approached the table, the gatekeeper had no need to check the list. As physician for the prison, Dr. Baly was always free to leave. His reputation however would remain forever locked within.

It was the winter of 1854 and the vast, brooding fortress of Millbank Prison loomed over the north bank of the Thames. It was not only the largest prison in England, but also one of the largest buildings in London, encompassing more than sixteen acres within its hexagonal walls. Oppressive in every way, a contemporary critic called it "the most successful realization, on the large scale, of the ugly in architecture."

In time the outer gate swung open and Baly walked toward the Thames, crossing the narrow field that had once been a moat surrounding the prison. A nauseating stench drifted upstream from the bone-crushing yards and the gasworks of Lambeth, but a far more worrisome cloud followed the man as he climbed into a waiting coach. Even the acrid yellow smoke that poured out of the gasworks and smothered the district every Friday would fade as he rode away from the prison, but his oppressive troubles would travel with him.

Five years earlier Dr. Baly had agreed to prepare a report for the Royal College of Physicians on the cholera epidemic of 1848–1849. With the full scientific and medical authority of the Royal College behind it, the report was to have been the definitive word on the cause of cholera. The college's cholera committee surveyed hundreds of physicians from throughout Great Britain and combed through mortality data to ensure they had a complete and accurate picture of the epidemic and its possible causes. The survey included a set of questions on local water supplies and John Snow had eagerly anticipated the "considerable amount of information" on drinking water and cholera that would soon be available to the medical community. That was 1849. Five years later the report had yet to be published.

For those five years, Dr. Baly had been under growing pressure to complete the report. The return of cholera early in the fall of 1853 increased the clamor for its release and forced Baly to finish it. By December the report was written, but his problems were far from over. The report had been intended to demonstrate the intellectual and scientific superiority of the Royal College as compared to the General Board of Health, which had issued its own report. If anything Baly had done the opposite, stretching the tolerance of the college toward the breaking point and reflecting a sense that the medical elite were at a loss when it came to protecting the public from cholera. Later in December, when Baly sat down to write the letter that would accompany the published report, he faced problems far more critical than a late assignment. Cholera was on the loose in Millbank Prison.

The letter, addressed to John Ayrton Paris, MD, DCL, FRS, President of the Royal College, consisted of a long list of excuses. First, Baly explained, reading through the survey responses required more time than expected. Then in 1850 as that task was completed, the reports from the General Board of Health (GBH) were released and, of course, the Royal College report could not be released until Baly had a chance to sift through the thoughts of the GBH. Shortly thereafter, he went on, he had learned that William Farr would be issuing a report on cholera mortality statistics and he wanted to incorporate that information in the Royal College's report. However when Farr's report came out in 1852, Dr. Baly concluded that he would need the raw data to draw his own conclusions. All these delays, he insisted, were unavoidable and the quality of the final report would demonstrate the necessity of the five years of preparation.

One could give Dr. Baly the benefit of the doubt. Cholera had confounded the world's best minds. Perhaps he had done his best to unravel the riddle. Perhaps each new piece of information was bringing him closer to the answer and the report could not have been completed any sooner. On the other hand, the reasons for the delay may well have lurked behind the thick stone walls of Millbank Prison.

When cholera struck London, the prisons and mental hospitals often stood out like islands amidst the sea of their surrounding neighborhoods. Because many of these institutions had their own water supply, cholera incidence rates could rise or fall sharply as one entered their confines. Inmates of the Hospital of St. Mary of Bethlehem (an institution for the mentally ill memorialized by the word *bedlam*) died from cholera at a rate that was one third that of the unconfined residents around them in Lambeth. The inmates of Queen's Prison also fared better than residents around them in the district of Southwark. Both the Hospital of St. Mary of Bethlehem and Queen's Prison had their own wells. The districts that contained them, Southwark and Lambeth, relied on piped water from the Thames.

The Millbank Prison had no well. Outside its gates a pipe jutted out into the murky waters of the Thames. Deep inside the prison, the inmates operated a pump that pulled water through this pipe into a large cistern. The cloudy water then passed through a filter, which gave it the appearance of clarity. When it came to removing *Vibrio cholerae*, however, appearances were deceiving. The inmates of Millbank Prison relied on this water to survive. In 1848 that reliance took a deadly toll. The epidemic of 1848–1849 had hardly begun before deaths from cholera filled the prison's graveyard to capacity.

It was John Snow who first made these observations. In October 1849 William Baly picked up a copy of the *London Medical Gazette* and discovered Snow's description of conditions at the Hospital of St. Mary of Bethlehem and Queen's Prison. Then came the shock of reading an indictment of the water supply at Millbank. Snow concluded not only that the water supply explained the prison's high rates of cholera, but that it explained the incidence of outbreaks of dysentery in the intervening years as well. The report never mentioned Baly by name, but it implied that the prison physician had presided over a public health disaster.

Snow had to be wrong. Baly had overseen the operation of the filter and assured himself that the filtered water was clear. Moreover the unique nature of Millbank Prison convinced him that cholera could not be spreading from prisoner to prisoner. Based on a system developed by Jeremy Bentham, the father of utilitarianism and the mentor of Edwin Chadwick, Millbank kept its inmates in almost complete isolation. The majority of prisoners spent most of the day working alone in their cells. Even on the occasions when they were together, they were forbidden to speak or even come close to one another. In Baly's mind, only an airborne agent could spread cholera throughout the prison.

For a thoughtful man like Baly, making Snow's truths go away was no simple task. He devoted one in every ten pages of the report to counterarguments and examples intended to refute Snow's theories. In 1853 however medical science often relied on anecdote, subjective observation, and "expert" opinion. Baly's report was no exception. Its consideration of drinking water depended on the opinions of physicians throughout the country. Its assessment of the water at

Hertford County Prison, for example, relied on the correspondence from a Dr. Davies who considered it to be "extremely sweet and pure." The Wakefield Lunatic Asylum also had water that was "quite pure" according to Dr. Wright, a local physician. The water consumed by children at Dronet's school was also "not contaminated" in the words of one Dr. Kite. All three institutions had suffered massive cholera outbreaks. Baly relied on the words of these three men as critical evidence of the errors in Snow's theory.

After twenty-three pages of similar examples, weak science, and flawed logic, Baly concluded that "It is not probable that in the case of Cholera the influence of water will ever be shown to consist in its serving as a vehicle for a poison generated in the bodies of those who had suffered from disease." His next sentence began with the word *but*. He simply couldn't bring himself to reject Snow completely. Perhaps "the poisonous matters which produce cholera ... are capable of increasing in foul water as well as foul air." "It is scarcely probable," he continued, "that water containing putrid matters in a state of solution or of suspension can be habitually swallowed without at least the risk of injury to the health." He then recapitulates the theory of William Farr that the poisonous matters might well be rising with the mists of evaporating water from the Thames. With his own Millbank prison perched on the banks of the Thames, Baly found a way to blame the river for its troubles. The miasma had emerged from the river and crept over the prison walls bringing cholera under its wing, an explanation that just happened to absolve him of responsibility.

As Baly left Millbank Prison on that winter day in 1854, even the pending publication of his report brought him little solace. Cholera had returned to London a third time and the inmates of Millbank continued to suffer from serious disease at a rate far above anything experienced in London's other prisons. Not only was this situation a public health crisis, but as Baly was recognized as one of the nation's leading authorities on prison hygiene, it posed a direct threat to his professional reputation. Unfortunately for Dr. Baly and his charges, the coming year would unleash a catastrophe within the walls of his prison.

John Snow read Baly's report with a cool, objective eye. Unfazed by its critique of his ideas, he made careful notes of its weaknesses and its strengths. Even though it bore the imprimatur of the Royal College, Snow was not rattled. He would respond in his own way and in his own time. Meanwhile he was immersed in a far more important project.

Epidemiology must count corpses to make a case. With the medical establishment continuing its efforts to explain away Snow's findings, it was not clear how many more dead bodies would be required to overcome their resistance. It was clear however that cholera would provide them and, even before it returned in 1853, John Snow was counting.

Ever since 1849 he had gathered whatever information he could on the London water companies and their relationship to cholera mortality during the first two epidemics. He had already demonstrated that districts using unfiltered water from the Thames tended to have higher rates of cholera mortality than districts with cleaner sources of water. His critics focused on inconsistencies in this pattern pointing toward high rates in districts with "clean" water or low rates in districts with dirty sources. Snow had explanations, but he would have preferred decisive evidence. By 1853 the water companies themselves would give it to him.

In 1852 The Metropolitan Water Act had given the water companies three years to move their water supplies away from the most contaminated reaches of the Thames. When cholera struck in 1853, the quality of the Thames water was in a freefall and little had been done to improve the public water supply. John Snow had been working on additional analyses of the role of the water companies in earlier epidemics, but the results fell short of what he sought. He needed irrefutable evidence. In the winter of 1854, even as he read Baly's report, he had already begun to believe he might have found it.

The decision by the owners of a single water company to move their intake before the deadline set the critical experiment in motion. Until 1853 the Lambeth waterworks had drawn their water from the Thames in the center of London, just opposite Hungerford Market. Then two years before the 1855 deadline, they finished work on a water pipeline that reached out to a new steam-powered water pump at Thames Ditton. The pump's intake pulled water from the river at a point more than ten miles upstream from Hungerford Market and well out of reach of London's sewage. In the moment that the old pump was shut down and the new pumps first sent water rushing into the Lambeth Company's maze of pipes, one of London's dirtiest water supplies became one of its cleanest.

The move by Lambeth meant that the Southwark and Vauxhall Company (S&V) was now providing the dirtiest water in the city. S&V drew their water from Battersea, just half a mile upstream from the Millbank Prison. In no hurry to improve what they believed to be a clean water supply, its owners continued to send an unsavory cocktail of sewage-tainted water to their customers. Both Lambeth and S&V supplied the south bank of the Thames. If Snow could show sharply lower rates of cholera in the districts served by Lambeth, he would have a new and powerful piece of evidence. That analysis would be anything but simple.

Snow's challenge resulted from a war that had taken place twenty years earlier. The war was an economic war among the water companies of London that began in 1834, when Parliament eliminated a clause that had prevented competition among water companies south of the Thames. The Lambeth Company soon found itself locked in a vicious battle for territory with the two other companies supplying water to the districts south of the river, Southwark and Vauxhall, which were separate companies at that time.

For three years corporate warfare raged through the streets of London. Lambeth invaded the district of Southwark at the same time workers from the Southwark and Vauxhall companies made incursions into Lambeth. In the fight for customers, the companies sent pipes coursing side by side beneath the same streets. They did not even hesitate to invade the same building. Above the street the battle was so intense that brawls erupted between work crews from the different companies.

Soon the Lambeth Company found itself fighting for its life. In every district it served, it was locked in combat with Southwark or Vauxhall. On all sides the cost of the war soon became prohibitive. In 1837 the companies called a truce and agreed to halt the destructive competition, but the snarl of pipes remained.

In 1853 this meant that John Snow could not find a single district served by Lambeth alone. Southwark and Vauxhall had merged in 1845 and every district served by Lambeth was also served by the newly formed S&V. This had mattered little in his analyses of earlier epidemics because S&V and Lambeth had water of similarly poor quality. Now it was critical.

Snow was sure that cholera rates would be lower in homes served by the improved water supply of Lambeth. Through the course of the epidemic, William Farr and his staff at the Office of the General Registrar had tabulated cholera deaths in London and published them in the *Weekly Returns*, but these reports only listed results by district. If the Lambeth Company had no district as its exclusive province, how could Snow isolate the salutary effect of its improved water supply?

The beginnings of an answer came late in November from Farr himself. Farr's work on elevation and cholera was seen as the most important study of cholera of the time. Farr however was far more open-minded than many of his sanitarian colleagues and was intrigued by the idea that contaminated water might contribute to these low-lying miasmas. So he decided to assemble a table that grouped the districts according to water supply and published the cholera mortality for these groupings in the *Weekly Returns*. He gave separate figures for the districts served by S&V alone and those served by both S&V and Lambeth.

Snow observed that districts served by both Lambeth and S&V had, on average, half the rate of death from cholera as compared with areas served by S&V alone. This suggested that cholera rates were lower for Lambeth customers, but was not conclusive. Snow needed a way to isolate the customers of the two companies. Once again he obtained key information from William Farr.

Farr was nothing if not thorough. Early in 1854 the General Registrar published a list of the 796 deaths from cholera during 1853 in London. In the meantime Snow had broken down the water supply at the subdistrict level rather than the district level. Of the thirty-three subdistricts, twelve relied only on S&V and sixteen relied on both companies. Three small subdistricts, he found, were served by the Lambeth Company alone.

Snow spread out a large map and worked his way through the 796 names on the list. He eliminated any deaths from outside the thirty-three subdistricts. As he sorted through the 374 deaths that remained, a startling picture emerged. Norwood, Streatham, and Dulwich, the three subdistricts served only by Lambeth, had 15,000 residents. None of them had died from cholera in 1853.

But Snow needed more. Even if subdistricts that relied on contaminated water from the Thames had higher rates of cholera, those areas inevitably differed with respect to something other than the water supply. They might have lower income, poorer housing, inadequate drainage, or more miasma-generating industries. Norwood, Streatham, and Dulwich were more rural in character than most of the other subdistricts south of the Thames. He could not be certain that better water quality from Lambeth had been the true cause of lower

cholera rates. Perhaps it was just the life-affirming air of the highlands. Today epidemiologists refer to this problem as confounding.
Snow had no word for it, but knew he must address it.

Snow began to believe that the evidence he wanted was not in
those three subdistricts served by Lambeth alone, but in the sixteen
subdistricts served by both companies. There, he suspected, an experiment had occurred "on the grandest scale." There the water companies had taken people living in the same neighborhoods, people of
similar means, living in similar houses and breathing the same air, and
divided them into two groups, "one group being supplied with water
containing the sewage of London, and, amongst it, whatever might
have come from the cholera patients, the other group having water
quite free from such impurity."

If he could demonstrate that in those subdistricts served by both
suppliers most deaths occurred among customers of S&V, his critics
would lose their most powerful argument. He simply needed to go
into the district as cholera cases occurred and identify the supplier of
water. John Snow had only one problem. By the time he recognized
what he had, winter had temporarily driven cholera from the streets of
London. He would need to wait until warm weather returned.

After battering London during the summer of 1853, cholera had pulled
back in the winter of 1854. But the disease had merely retreated, not
surrendered. It waited through the winter, holding on in isolated
pockets, until the water warmed in the sun.

When the waters of the English Channel yield to the tireless pull
of the moon, the tide rushes into the mouth of the Thames, halting
and then reversing the river's flow. Climbing upstream, the ocean current runs past Southend and Gravesend, past Purfleet and Frith and on
into London. In the summer of 1854, when the river was low, a salty
tongue of seawater reached through London to Battersea and licked at
the intake of the S&V waterworks.

At the same time, the Port of London bustled with ships. Many had come from the ports of Europe where cholera had smoldered through the winter. The warmth of summer had breathed life into the embers of the epidemic and by the end of June, the coasts of the Continent were ablaze with cholera. When the sailors who took refuge in the sordid quarters that surrounded the docks of these unfortunate cities climbed aboard their ships to sail to England, cholera sailed with them. When they arrived in London, these ships dumped their wastewater into the Thames. Soon the river was alive with the bacteria brought by afflicted seamen. By July the pathogens had ridden the tides to Battersea and cholera exploded again.

As cholera renewed its assault, John Snow went back to work. The heat of mid-August found him walking the streets of Kennington with a satchel and a list of addresses provided by William Farr. Over the winter he had developed a plan that he would implement if cholera returned. Kennington comprised two subdistricts that both relied on the two water companies. Removed from the river and any purported miasma and inhabited by a broad mix of social classes, it seemed an ideal choice for his initial study.

John Snow had spent the day knocking on doors to talk to the families and friends of cholera victims. At each home he had expressed his condolences and asked the name of the water supplier for the house. More often than not, the response was a puzzled face. He waited as homeowners disappeared to search for a receipt from the water company. Frequently they returned empty handed. Again and again renters confessed their ignorance with respect to their water supply. The intermingling of the water supplies, it seemed, was so complete that it might defy his efforts to record the progress of this natural experiment.

John Snow had grown adept at navigating the ignorance of others. When he encountered uncertainty about the water supply, he opened his satchel, removed a vial and asked for a sample of tap water. He carefully labeled the vial and returned it to the satchel. If the building's residents could not reveal the source, perhaps he could find a way to make the water talk.

At first John Snow was not sure how he would sort out this puzzle. He brought water from each supplier to his laboratory to test, hoping he could find an answer. He had already noticed that when he held the vials of water up to the light, that he could see fine particles swirling in the S&V water, but this was not enough. The particles might settle out if the water had sat in a cistern. On a given day, they might be less prevalent. Even the Lambeth water could become cloudy if it had sat in a dirty cistern or had run through dirty pipes. John Snow needed certainty.

After five years of looking for the data that would prove water's capacity to kill, he sensed he had found the critical evidence. He could not let himself be thwarted by something so simple as the incomplete knowledge of the study subjects. But if he could not find a way to distinguish between the different water supplies, the results of that experiment would remain unrecorded. Finally John Snow had an idea.

That long tongue of seawater carried the answer to his problem. Salt. Snow took a gallon of water from each supplier to his laboratory. He pulled a bottle of silver nitrate solution from the shelf and added an equal amount to each sample. As he did so, a cloud formed in the water. The silver was combining with the chloride from the salt to form insoluble particles of silver chloride. Today high school students routinely perform this experiment, but in 1854 this was state-of-the-art analytical chemistry. When he dried the crystals from each sample, his problem evaporated as well. The S&V water, being closer to the mouth of the Thames and the tidal flows upstream, contained forty times as much salt as the Lambeth water from Thames Ditton. If he performed a similar test on his samples, he would know the source of their water with absolute certainty.

Forty-four times John Snow knocked on the doors of cholera victims in Kennington. Forty-two times he obtained a receipt or a water sample. (Two homes relied on pump wells for their water.) As he tested each water sample for salt and began to accumulate data, the results were astonishing. Ninety percent of the cholera victims got their water from S&V.

In the course of his walking tour from home to home, he had recognized something else remarkable. Nothing about the homes he visited allowed him to predict the water supply. There seemed to be no pattern to the distribution system. He had stumbled on a population that had received either water contaminated with sewage or pure water almost at random. The customers of the different water suppliers differed only with respect to the water they consumed. Otherwise they were as similar as two cages of laboratory guinea pigs. The war of the water companies had eliminated confounding. John Snow had found his *experimentum crucis.*

But he had only looked at a part of the experiment. He had looked at two subdistricts over a few weeks of the epidemic. Chance had performed this experiment on more than 300,000 people and would continue to do so as the death toll mushroomed. Such an important experiment needed to be recorded in full, but such an undertaking was far beyond his means. He needed help, but where could he go?

Snow could think of only one place. Only one group in England had the resources to gather this critical data. But that group was lead by an ardent sanitarian. He might well scoff at Dr. Snow and his wild ideas about water. Snow had to take that chance.

John Snow rushed to the Office of the General Registrar to talk to William Farr. Farr had pioneered the collection and use of data on death and its causes and had given Great Britain the world's best vital statistics registry. No one respected the power of these data more than Farr. Despite Farr's belief that cholera emerged from low-lying miasmas rather than water, he saw in Snow a kindred spirit. The two men had helped found the London Epidemiological Society and saw each other regularly at its meetings. When he saw Snow's data, he too was astonished.

In response Farr himself suggested that the registrars of the south districts of London should determine the water supply for all homes in which a death occurred from cholera. After years of solitary research, Snow finally had an ally.

Still Farr's order would only cover deaths that occurred after August 26. Snow had studied deaths through August 12 and only in a portion of the subdistricts supplied by the two companies. Snow was determined to gather all possible data. He would need to expand his study to include all sixteen subdistricts and two additional weeks during which cholera mortality had risen explosively. That meant he would need to visit the homes of more than a thousand cholera victims. He would need help. And time.

The earnings from his burgeoning medical practice allowed him to hire John Joseph Whiting, an apothecary, to assist in the study. For the first and only time in his career, John Snow had a research assistant. He assigned Whiting to visit homes in the districts served only by S&V. In that way Whiting would only need to determine whether or not the victim had used piped water, well water, or water from some other source such as tidal ditches. Snow would take on the more difficult task of determining the water supply used by victims in subdistricts served by both suppliers.

He had set aside the first two weeks of September to complete his task. And so Dr. John Snow, anesthetist to the queen, prepared to walk the streets of South London, visiting hundreds of homes and collecting and analyzing vast numbers of water samples to determine the source of their drinking water. Just as he began this daunting task, cholera came to his doorstep.

5

THE DOCTOR,
THE PRIEST, AND
THE OUTBREAK
AT GOLDEN SQUARE

As an oppressive blast of late summer heat bore down on the estates of Hampstead, just north of London, Susannah Eley, a wealthy widow, was enjoying a visit from her niece. For almost the whole of August, England had baked under clear, motionless skies. What breeze there was carried in a blanket of thick, humid air laced with the fetid breath of London. As that last, cloudless day of the month wore on, the two ladies sipped from their water glasses in the sweltering heat.

They found nothing disagreeable in the taste of the water. It gave off no foul odor, nothing to suggest the presence of a disease-causing miasma. If either of them had looked closely, she might have seen a few fine white particles drifting in her glass, but in 1854 this was no cause for alarm. What possible harm could come from particles they could barely see? They had no reason to believe that death could come in such a small package.

The same heat that made the widow Eley and her niece uncomfortable made London's inner city unbearable. Day after day the sun had slowly roasted the city's noxious accumulation of human and animal waste. Each neighborhood had its own special topography of odor that rose and fell with the temperature. On a cool day, the terrain was challenging, but negotiable. In the heat of summer, the city's residents struggled to find a path through these mountains of stench.

For Londoners who lived or worked among the dense jumble of residence and commerce between Golden Square and Soho Square on London's East End this assault on the senses was unrelenting. At the street level, every manner of business and industry from breweries to slaughterhouses produced a full spectrum of offense. To this mixture stables, decrepit privies, and ancient cesspools added their own offerings, sending an unrelenting stink steadily up toward the homes of the working poor who crowded into tiers of one-room apartments above the street.

When the heat was less intense or the air less still, those who lived and worked on these dark and narrow streets might find a mouthful of untainted oxygen in the open expanse of Golden Square or on the few wide streets such as Marlborough or Broad, but they provided no such haven on that Thursday. As the heat grew on that last day of August, the vile reek of the privy outside the Newcastle-on-Tyne Pub at the corner of Cambridge and Broad laid siege to the four-story tenements next door. The residents crowded into the rooms above the boot tree maker on 40 Broad Street could either close their windows and swelter in the stultifying heat or open them and surrender to the putrid assault.

In one of those rooms, Susan Lewis, the wife of a policeman, anxiously prepared a bottle of rice meal and milk. In a world that did not understand sanitation, bottle-feeding often meant disease and even death for an infant, but Susan Lewis had no choice. An illness late in pregnancy had left her unable to breast-feed her infant daughter. The baby had now fallen seriously ill for the third time in her five short months of life. The attack of diarrhea, which had begun just four days earlier, had subsided, but it had left her weak with no appetite. Mrs. Lewis tried to stir her baby, offering the bottle and hoping she would eat. Instead the frail infant lay passively, taking nothing. Three years earlier Susan Lewis had lost her son before he reached his first birthday. Now the young mother feared the worst.

Two doors down workers at the Eley Brothers percussion cap factory replenished two large cisterns with water from the pump just outside their doors on Broad Street. In the heat the owners made sure to refresh their workers' water supply regularly. They had recently noticed that the water tended to develop an offensive odor after sitting for more than two days, so they were particularly diligent in this task.

That morning workers at the factory had also filled a large bottle with water from the pump and packed it on a cart bound for the West End. For reasons lost to history, the Eley brothers' mother preferred the water from the pump to that from wells far closer to her home. Perhaps she requested water from Broad Street out of some sort of affection for her late husband who had owned the factory until his death. Perhaps she simply preferred the taste of its water.

Whatever the reason Susannah Eley's sons sent a cart laden with drinking water from the Broad Street Pump on its routine trip. Several times each week, the cart would make its way through the crowded streets of London, past the farms that surrounded the great city and out to Hampstead, a distance of four miles, to deliver a large bottle of water and, in the late summer of 1854, cholera.

As night fell on the lingering heat of that cloudless summer Thursday, the widow Eley and her niece drank the water from Broad Street, oblivious to the disaster that lay ahead. They could not have imagined that each of the minuscule particles suspended in the innocence of drinking water contained millions of deadly comma-shaped bacteria, the telltale form of *Vibrio cholerae*. That night the strong acids in the widow's stomach destroyed almost all these microscopic invaders and might have saved her had they not been so numerous. The massive dose in each glass of water ensured that some of the bacteria would find their way through her stomach to the haven of her small intestine.

The hydrochloric acid that had dissolved her evening meal packed the corrosive power of battery acid. In addition to destroying most of the bacteria, it would have eaten through the walls of her stomach were it not for the thick protective slime secreted from its lining. This coating worked well for the crude operations of the stomach. The small intestine had a far more intricate task. It would need to break down the nutrients in the slurry of food passing through its thirty-foot length and transport them molecule by molecule into the widow's bloodstream. Any protective layer would make this impossible. Instead specialized cells released just enough bicarbonate to neutralize the acidic mixture flowing from her stomach to protect the delicate lining of her small intestine. That night the bicarbonate also granted a reprieve to the cholera.

As the night wore on, the few surviving bacteria had already begun to reproduce inside her. When she awoke their numbers were still so small that Susannah Eley had no notion of the danger at hand. Back on Broad Street, however, disaster had already begun to rear its head.

Homes throughout the neighborhood had spent a night in sheer terror as family members plummeted into the abyss of cholera. On the morning of Friday, September 1, as the people of Soho stepped outside, they found that a cooling breeze had arrived from the northeast. The heat wave had faded, but something far worse had emerged in the dark of the night. Word of the horror buzzed through the neighborhood. Almost everyone, it seemed, knew someone who was dying.

No one understood the scope of the disaster better than Reverend Henry Whitehead, the priest of St. Luke's Church, which stood just one block from Broad Street. Throughout the previous evening and into the night he had walked the streets of his parish in long, flowing robes ministering to the afflicted and their families. For many the appearance of his broad, friendly face gave them a moment of reassurance in an otherwise desperate night. When he finally made his way home through the dense mix of mist and coal smoke to the small apartment that he shared with his brother, he sensed he was facing a devil unlike any he had encountered in his twenty-nine years. He would spend much of the next day delivering last rites to the dying.

Word of the emerging tragedy had yet to reach the affluent district just across Regent Street where John Snow had immersed himself in a study of the London water supply. His office was filled with maps, death records, and the data he had assembled over the preceding year on the relationship between water supply and cholera deaths in an area south of the Thames. He had recently cut back on his clinical work so he could devote even more time to his *experimentum crucis*. So as the morning of Friday, September 1, arrived, he was simply relieved that the heat wave had broken.

After six years of struggling to convince the medical establishment that drinking water could spread cholera, John Snow felt he was finally closing in on the proof that might muffle his critics. The ground itself would need to shake to divert his attention from the task at hand. He would soon discover that an epidemiological earthquake like no other had its epicenter on the north side of the Thames, just a short walk from his front step.

But as the sun rose on that Friday morning, John Snow had no notion of the horrifying turn of events so close to the desk where he sat piecing together the story of the London water supply. He could not see the first black, windowless carriage as it appeared from behind the curtain of fog that the cool air had draped across Broad Street. He could not hear the horses protest as their driver, also dressed in black, pulled at their reins, bringing the carriage to a stop in front of a four-story building. Two men climbed down and entered the front door with a stretcher. They climbed to an apartment marked by a row of shuttered windows. Minutes later they emerged carrying a draped, lifeless figure. Before they had finished loading the corpse into the back of their carriage, the sound of hoofbeats and steel-rimmed wheels signaled the arrival of another crew with the same grim mission.

Over the course of that day, a fleet of hearses rolled steadily through the district. Again and again these faceless wagons came to collect their tragic cargo. More than sixty people living in the area around Broad Street died on that Friday. As each hearse arrived, neighbors watched, quietly registering the address of the deceased and wondering where the next would stop.

But the worst was yet to come. Throughout the day the blue death roamed the district, visiting one home after another, selecting its next victims with an apparent randomness that was both cruel and terrifying. As word spread of each new case, fear grew palpable. Those who rented furnished rooms and did not have a friend or relative in need of their care packed up their belongings to seek refuge elsewhere. Before Friday found its end, 143 more people were fatally ill. Still the ravages of cholera were just beginning.

Of those to whom cholera had laid siege on that Friday, all but two lived within a few blocks of the Broad Street pump. To reach these last victims, however, the disease had traveled to the comfortable homes of Hampstead. There, far from her deceased husband's Broad Street factory, Susannah Eley began to feel a vague discomfort.

The bacteria in her small intestine had spent much of the day doubling and redoubling. As the *Vibrio cholerae* multiplied, they busily manufactured a deadly poison. The toxin targeted the switches in the lining of the intestine that controlled the flow of bicarbonate, jamming them into the "on" position. Susannah Eley could not notice the initial trickle of fluid that seeped into her gut. But as the numbers of bacteria grew, vast quantities of bicarbonate began to flow into the widow's small intestine.

By hacking into the signal pathways of her digestive system, the bacteria created a surging torrent of acid-neutral fluid, full of nutrients in which they could grow and reproduce. What had been the machinery of digestion became a factory for the production of billions upon billions of pathogens that rode the flood of bicarbonate out into the world in search of other victims. It was not the bacteria, but the river they created that would kill her.

As the first day of September tumbled into darkness, cholera tightened its grip on Broad Street. Early on Saturday the parade of hearses began again. Throughout the day they arrived with increasing frequency, each one departing laden with its grim payload. Late in the afternoon, one arrived at 40 Broad Street. It carried off a tiny corpse and left Susan Lewis consumed with grief over the death of her second child.

Even in a city that had learned to live with cholera, a disaster of this intensity attracted notice. By Sunday, September 3, word had spread through London. When the news reached the office on Sackville Street where John Snow was untangling London's complex network of water supplies, it found the one person who could understand what had happened.

In the mind of Snow, only drinking water could cause such a sudden explosion of cholera. He also knew that two water companies supplied piped water to the area affected by the outbreak; the Grand Junction Company supplied the western portion near Golden Square and the New River Company supplied the eastern portion near Soho Square. Not only did both companies have relatively clean sources of water, they drew from entirely different rivers. The simultaneous contamination of both sources defied probability. More important, Snow knew that anything involving these companies would have affected a much larger area than the highly localized disaster at hand.

As he gathered his hat and stepped out into the fading heat of summer, John Snow felt certain that the "morbid matter" that caused cholera had made its way from a victim of the disease and into some single, common source of water. He reasoned that a contaminated public well must have given rise to the outbreak.

As usual the Sunday evening traffic was light as he made his way across Regent Street. Not only did John Snow understand cholera, he knew the Broad Street neighborhood as well as anyone. His first two residences in London were just a few blocks east of Broad Street and now he lived in a comfortable home a few blocks to its west. There was only one pump that could have caused the pattern of death that his medical friends had described to him and he was walking straight toward it.

His footsteps echoed across the cobblestones as he walked through Golden Square and turned onto Silver Street. Even for a Sunday, this gritty, commercial neighborhood was unusually quiet and empty. He continued on past the National School and onto a wide street that was just over four blocks long. Passing beneath the watchful eye of the stone lion perched above the Lion Brewery, he could see the yellow pestilence flags that hung limply from the lampposts. Throughout the deserted street, the shuttered windows of mourning marked the homes of those who had already died. Halfway down the street, he stopped. He carefully removed a small bottle from his pocket, lifted the handle of the Broad Street pump, and held it under the spout.

Even Snow had to rely primarily on his senses to test the water. He examined the contents of the bottle looking for any hint of contamination. He saw and smelled nothing to suggest the presence of organic matter in the water. His years of scientific training had taught him the value of skepticism, particularly in the evaluation of one's own, best ideas. He left unconvinced that he had found the source of the outbreak.

He continued on through the neighborhood, determined to find the water pump that cholera had used to launch its attack. At the Warwick

Street pump, he saw small white, flocculent particles in the water. He found similar impurities in the water at Bridle Lane. The dirtiest water came from the pump on Marlborough Street, but passersby informed him that this was well known in the area. Most people, he learned, preferred to gather their water from the pump on Broad Street.

When John Snow returned home that night, he remained convinced that a water pump had caused the ongoing disaster. The pump on Broad Street remained the prime suspect. He was equally certain that finding the exact cause and demonstrating its role to a deeply skeptical audience would require far more work. No one in London shared John Snow's concern about the water from the Broad Street pump. Those who remained in the area continued to collect its water, just as they always had.

As the last hour of that tragic weekend slipped away, Reverend Henry Whitehead sat down to rest. This had been a weekend like no other in his life. All his waking hours had been spent as an observer at death's door trying to bring some measure of comfort to the dying and the desperate. Drained of his physical, emotional, and spiritual energy, he sought to soothe his own soul with a glass of brandy. He diluted it with water he had drawn from the Broad Street pump.

John Snow's continued inquiries led him with increasing certainty to a single explanation for the pattern he saw in the mounting cholera deaths. On Monday, after providing chloroform for a tooth extraction, he returned to the Broad Street pump. Even if he could not see it, smell it, or taste it, cholera's cause must be hiding in the well. This time, he took several samples to test.

Using one of the few reliable tests available to him, Snow added a few grains of silver nitrate to the sample. He shook it and the water grew cloudy with crystals of silver chloride. The fine grains sank in the water and accumulated on the bottom of the flask. Even before he weighed them, experience told Snow that the amount of chloride in the water was high. Chloride marked the presence of sodium chloride or salt. Since pure well water should contain very little salt, something must be contaminating the well.

Snow took another sample to Dr. Arthur Hill Hassall, one of London's most eminent microscopists and author of an authoritative book on the minuscule inhabitants of water. Hassall reported finding organic matter and oval "animalcules" in the sample. These may well have included *Vibrio cholerae*, but at the time Snow had no way of knowing this. Instead he only took this to be evidence of organic contamination, all the evidence he needed to continue to focus on the Broad Street pump. Somewhere in those contaminants, he reasoned, cholera's demon was hiding.

The members of the medical establishment would not see the evidence the same way. After all hadn't Snow's own examination shown that other pumps were also contaminated, perhaps more contaminated than the one at Broad Street? Snow's theory required them to believe not only that undetected contaminants could kill, but also that this would happen only when a special variety of these invisible killers was present. At a time when microbiology did not exist as a science, he would need the tools of epidemiology to make his case. But the tools he needed did not exist, would not exist, until John Snow invented them.

On Tuesday, at his first opportunity, he set out across Piccadilly Circus and then through the maze of London's streets until he reached the Strand, a grand promenade in front of the imposing buildings that lined the north bank of the Thames. There he entered Somerset House, a vast stone edifice that was home to the government's leading societies of science and art. Deep in its marbled bowels, he made his way to the Office of the General Registrar, where William Farr and his employees would be assembling the official record of cholera deaths for the preceding week.

That same morning the British Government had taken official notice of the outbreak. Sir Benjamin Hall, president of the General Board of Health, toured the scene of the disaster. Three-quarters of the area's residents had fled. Those that remained and were able clustered around Dr. Hall in the hope that the attention of the authorities would bring them some relief. Tall and handsome, Hall's striking figure earned him a nickname he would one day share with the largest clock in London, Big Ben. On this grim day, there was no missing him and his entourage as they walked the streets around the Broad Street pump. As they moved the hearses were busy again. On that day alone, forty-five people would die.

The next day as he monitored his anesthetized patients, John Snow's mind buzzed with thoughts of the unfolding outbreak. The general registrar had given him a list of all cholera deaths in London for the period from Thursday, August 31, through Saturday, September 2, the first weekend of the outbreak. Of the eighty-nine people on the list all but six had died within a few blocks of Broad Street.

As soon as he finished his clinical work, he returned to Broad Street and began to work his way down the list. At each household where someone had died of cholera, he would express his condolences and in a quiet, husky voice, he would inquire as to the habits of the deceased. Eighty-three times, he asked if cholera's victim had used the water from the Broad Street pump.

Within twenty-four hours, he had eighty-three answers and a clear pattern had emerged. Seventy-three of the victims had lived closer to the Broad Street pump than any other and every one of them routinely drank its water. Of the ten who lived closer to other pumps, five had preferred the water from the Broad Street pump and always requested it and three others were children who went to school on Broad Street and routinely drank from the pump. Out of all the cholera deaths in the area, only two of the deceased drew their water from other wells.

On the following Thursday, one week after the outbreak began, the Board of Guardians of St. James Parish assembled in an emergency meeting. Hundreds had already died in their small parish and the death toll continued to mount. By virtue of England's poor law, the board was charged with overseeing care for the indigent in an area of northeast London that included Broad Street and Golden Square. As they met to discuss the crisis in their midst, a stranger arrived at the Vestry Hall requesting an audience.

A balding man with intense, deep-set eyes entered the room and introduced himself as Dr. John Snow. He had spoken in the building before at the meetings of the Westminster Medical Society, but never had his message been so urgent. Standing in the elegant boardroom with floor to ceiling windows overlooking Piccadilly, Snow laid out his indictment of the pump and implored the board to close it. The board members listened as he described his investigation of the eighty-three cholera deaths and his indictment of the Broad Street pump. They doubted that drinking water could have caused such an epidemic, but ready to try anything, they relented. After all, anyone who could had fled in the face of the outbreak. When those still in the neighborhood came for water the next day, they found that the Broad Street pump had no handle. For the first time in history, a governing authority had taken action intended to halt an outbreak of waterborne disease.

In the fields of epidemiology and public health, removing the pump handle has become the stuff of legend, but even Snow himself was not convinced that doing so had saved any lives. Even though thirty-two people died of cholera on that Thursday, the outbreak was in decline and might well have abated without the closure of the pump.

Ultimately 623 people in this small neighborhood died of cholera in just over a week. Hundreds if not thousands more had contracted the disease, but managed to escape with their lives. London had not seen an outbreak of such focused ferocity since the darkest days of the bubonic plague. Most of the survivors had left the area, turning the normally busy commercial district into a ghost town.

Some had nowhere to go. In the wake of the disaster, a struggling intellectual by the name of Karl Marx sat in a sparsely furnished Soho apartment and recorded his impressions in a letter to Friedrich Engels.

> [T]he total absence of money is the more horrible—quite apart from the fact that family wants do not cease—as Soho is a choice district for cholera, the MOB is dropping dead right and left (e.g. an average of 3 per house in Broad Street) and victuals are the best defense against the beastly thing.

As cholera burned itself out, tragedy gave way to mystery. The outbreak had ended, but the investigation of its cause was just beginning.

By the middle of September, the disaster had spawned no fewer than four studies, each relying on its own particular brand of epidemiology and sending investigators scurrying through the neighborhood in search of cholera's cause. The most prominent among these came from the General Board of Health, which had charged three of its members with investigating a long list of concerns including atmospheric conditions, ventilation, the presence of nuisances and noxious trades, bad smells, privies and cesspools, the state of basements, and the quantity and quality of the water supply in the affected area. The fact that the water supply was included at all may have been a begrudging nod to the theories of Dr. Snow, but its place at the bottom of the list and the nature of the remainder of their tasks leaves little doubt that they were on a mission to find the miasma that had unleashed the epidemic.

As the Board of Health team sniffed its way through the neighborhood, the worst stink of summer had already diminished. Instead the acrid odor of lime overwhelmed the district. As part of a daily cleansing ritual, the Board of Guardians had ordered workers to coat the streets where cholera had struck with the heavy white powder. The blackened figures of mourning survivors cast somber shadows on what appeared at first glance to be a late summer snowfall. It seemed as if cholera had drained the color from Broad Street. Day after day John Snow made his way around snow white puddles to the homes of the mourners and asked about drinking water.

Snow had been a regular visitor to the Office of the General Registrar since the outbreak began and his list of names had grown steadily as the death toll continued to rise. In addition he had recognized that the official tally was missing many deaths of people who had not died in their homes. Middlesex Hospital, five blocks from Broad Street, provided the only available care to the destitute who lived solitary, transient lives clinging to the lowest rungs of society's ladder. Years later Florence Nightingale would recall her experience as a young nurse there during the first weekend of the outbreak. For three days she had worked without sleep as a steady stream of dying prostitutes and their fellow denizens of the street arrived with the icy hand of cholera around their throats. For days she could do nothing more than comfort them and watch them die.

Others were lost to the system because they had suffered the fatal misfortune of consuming the water from the pump before leaving to die elsewhere. At a chance meeting with Dr. Charles Frasier, the only physician on the team sent by the General Board of Health, Snow learned of two such cases. The first involved a man from Brighton who had come to care for his brother, but arrived to find he had died. The man saw no reason to linger in the presence of cholera, but he was hungry from his journey and had a long trip home. His sister-in-law prepared him a quick meal of rump steak. He washed it down with a tumbler of brandy and water from the Broad Street pump. After twenty minutes he was back on the road to Brighton carrying a lethal dose of cholera.

The second case involved the mother of three sons, all of whom worked on Broad Street. For Snow it was the exception that proved the rule. He had already spent considerable time talking to the brothers as they ran a business where sixteen workers had died of cholera. The business was Eley Brothers' Percussion Cap Factory and one of the three brothers told him of the regular deliveries of water to their now deceased mother. After following the path of the cart out to the home of Susannah Eley in Hampstead, Snow learned that the widow's ill-fated niece had died on the same day as her aunt, miles away in Islington. Cholera was a stranger to both of these districts. The only thing that connected these isolated and simultaneous cases was a bottle of water from Broad Street.

Snow's conclusions may seem painfully obvious to us, but we have the advantage of knowing he was right. The rest of London was looking elsewhere for the answers. In the words of Dr. Edwin Lankester, a prominent physician and member of the Board of Guardians who had listened to Snow's plea to remove the pump handle, "not a member of his own profession, not an individual in the parish believed that Snow was right."

As the neighborhood sought to regain its equilibrium, many of the survivors developed their own theory as to what had happened. Almost two hundred years earlier, William Earl of Craven had lived in an estate on Drury Lane, not far from what would become Broad Street. As the black plague ravaged the city, he had built several dozen pest houses in an area that included the west end of the future Broad Street. Victims of the plague, instant pariahs in a world that fled at the sight of their sore-ridden bodies, would come to this pest field to find refuge and in most cases to die. Their corpses accumulated in vast pits near the field. Many survivors of the cholera outbreak became convinced that recent excavations for new sewer lines had disturbed the pest pits and unleashed the remnants of this long-buried evil on the neighborhood.

This belief gained such credibility in the neighborhood that Karl Marx wrote in another letter to Friedrich Engels,

The cholera epidemic, now much abated, is said to have been particularly severe in our district because the sewers made in

June, July, and August were driven through the pits where those
who died of the plague in 1668 ... were buried.

Spurred on in part by this concern, the Commission of the Sewers
sent an engineer by the name of Edmund Cooper to investigate the
condition of the sewers in the neighborhood. The commission however
had no intention of accepting blame for this disaster. His report, issued
after just two weeks of peering down gully holes, concluded that bad
smells had indeed caused the outbreak, but went on to state that the
smells had come from within the houses. The sewers, he affirmed, were
in good shape and could not have played a role.

One evening late in September, Reverend Henry Whitehead paid one
of his frequent visits to the oil shop on Broad Street. Outside the shop
a large barrel labeled CHLORIDE OF LIME in bold block letters
offered its contents to passersby. A few minutes later he left carrying
a large tin of kerosene. He would be staying up far into the night,
working by the warm light of a flickering lamp to put the finishing
touches on his own study of the outbreak that had struck at the heart
of his congregation. Within a few weeks, he had issued his report,
The Cholera in Berwick Street. It made no mention of the Broad Street
pump.

As a priest of St. James Parish, he knew of Snow's theory about the
pump, but was convinced that Snow was wrong. After all Whitehead
had drunk from the pump himself with no apparent ill effects. (He
could not have known that brandy at high concentrations could protect
him from contaminated water. We can suspect that, in the midst of
the outbreak, he did not add much water to his brandy.) Furthermore
he had been at the homes of three cholera victims who had consumed
quart after quart of water from the pump during their illness and had
recovered fully from the deadly disease. Whitehead boasted to a medical friend that he understood the dynamics of the outbreak far better

than Snow and could prove the irrelevance of the pump if given the time. He would soon be given the opportunity to do just that.

As the fall wore on, the community had no real answers as to the cause of the outbreak. The General Board of Health had finished gathering data, but there was no sign of a report. If, as expected, the report blamed uncontrolled odors within the neighborhood, it would represent an implicit indictment of the Board of Guardians and by extension the Vestry of St. James, since they were responsible for the public health of the district. On November 23, 1854, almost three months after the start of the outbreak, the vestry commissioned its own study.

Epidemiology is deceptively difficult. The committee assembled by the Board of Guardians discovered this after a questionnaire sent to every home in the neighborhood produced no useful information. The board then summoned Dr. Snow for his input, as he was already a respected epidemiologist regardless of his views on water and cholera. After a meeting on December 12, the Board of Guardians added eight new members to the committee. The additions included Dr. Snow and the Reverend Henry Whitehead. This would be the beginning of a remarkable relationship.

Whitehead made no secret of his belief that Snow was wrong and his determination to prove him so. He routinely worked until 4 A.M. assembling data. He visited every home on Broad Street, many of them several times, to ask in detail about the outbreak. He was convinced that he had identified a key flaw in Snow's work and in many ways he had.

Snow had looked almost exclusively at those who had died from cholera and asked about their exposure to water from the pump. However he did not investigate those people who did not die. What if everyone in the area drank from the Broad Street pump? In that case the fact that most of the victims had used the pump would mean nothing more than that they had resided in the neighborhood.

Despite their fundamental disagreement, Snow and Whitehead developed a growing respect for one another as the work of the committee progressed. They both worked tirelessly on their study of cholera and were both scrupulously objective in their review of the data. In January, when Snow published a revised and substantially expanded version of his monograph, *On the Mode of Communication of Cholera*, he gave a copy to Whitehead. The most prominent addition to the new version was an exhaustive account of his study of cholera rates among the London water companies. Snow viewed this as his masterwork, the definitive proof his critics had been demanding. He had also included a substantial section on his study of the Broad Street pump.

Upon receiving the slender volume, Whitehead read the section on the Broad Street outbreak with great interest. He pored over it, looking for two critical numbers, the rate of cholera deaths among those around Broad Street who drank from the pump and the rate of cholera deaths among those who did *not* drink from the pump. If the rates were similar, as Whitehead expected, it would confirm his belief that the pump water had not caused the outbreak. Snow had not included these key numbers, so Whitehead decided to generate them himself.

Over the next months, Whitehead worked feverishly to amass the data needed to calculate the death rates. He focused exclusively on Broad Street since this would leave no ambiguity about the proximity of the pump. Well-liked and highly respected among local residents, he visited homes again and again until he was certain as to their source of drinking water during the days surrounding the outbreak.

On one such visit, he sat with a man and woman whose adult daughter had died of cholera. The husband sat quietly as his wife told Whitehead that they all drank exclusively from their cistern, which stored piped water. Whitehead asked specifically what they drank on September 1, the day before their daughter died. She remembered her husband and daughter drinking gin and water that evening and assured Whitehead that the water was from the cistern. Unlike any of the others investigating the outbreak, Whitehead knew the people he interviewed. Knowing that the husband was almost deaf, he asked in a loud voice if his wife's recollections were accurate. No, the old man informed the priest, he and his daughter had used water from the Broad Street pump on that evening. He boiled his water, preferring a hot drink. His daughter drank it cold. Twenty-four hours later, she was dead.

That spring the General Board of Health finally issued their report on the outbreak. They put little stock in Snow's contention that "the real cause of whatever was peculiar in the case lay in the general use of one particular well, situate [*sic*] at Broad Street in the middle of the district, and having (it was imagined) its waters contaminated by the rice-water evacuations of cholera patients. After careful inquiry," the report concluded, "we see no reason to adopt this belief." Instead they rambled on for more than three hundred pages in a convoluted effort to explain how unique atmospheric conditions conspired with the inadequate sanitation in the district to ignite the epidemic.

By then however Whitehead's late nights parsing data were leading him to an entirely different conclusion. After interviewing more than four hundred residents of Broad Street, he determined that those who drank from the Broad Street pump in the days surrounding the outbreak were nine times more likely to have died from cholera than those who used water from other sources. Epidemiologists now refer to these comparison rates in exposed and unexposed groups as a relative risk, but Whitehead had no such term. He only knew that his endless hours of work to establish the innocence of the pump had done the opposite. He had developed compelling evidence that Snow was right.

Still no one could find the source of the contamination that had caused the cholera. Where was the index case, the case that preceded all others? Then on March 27, 1855, long after others had closed their casebooks on the outbreak, Henry Whitehead was scrutinizing death records when something caught his eye. An infant had died of diarrhea on September 2 after an illness of four days. That would put the onset of illness at August 29, two days before the start of the outbreak, roughly the time it takes to contract cholera. The line that stopped Whitehead in his tracks was the address, 40 Broad Street, right in front of the pump.

He rushed off to speak to the mother of the baby. He found the grieving widow dressed in black in the shuttered darkness of her room on 40 Broad Street. In addition to losing her child to diarrhea, she had lost her husband to cholera. Her name was Susan Lewis. She described her desperation on August 29, as she struggled to attend to her daughter who suffered from profuse diarrhea. She had worked constantly to keep her clean, returning time and again to the basement where she washed the soiled diapers. For two days she had poured bucket after bucket of wash water into the sink, which drained into the building's decaying cesspool.

Next Whitehead consulted Dr. Rogers, the physician who had cared for the dying baby. Rogers felt that the symptoms were not at all typical for cholera and believed that this was some other ailment. Undeterred Whitehead convinced the Cholera Inquiry Committee of St. James to commission the excavation of both the well and the cesspool for 40 Broad Street. When the subsurface was laid naked, it showed a clear muddy trail. Each bucket of water from the dying infant had sent a plume of contaminated water into the soil around the cesspool. From there it had seeped across three feet of saturated soil, and down into the well for the Broad Street pump.

In mid-nineteenth-century London, one in eight children never reached the age of five. The plight of this infant would have received little notice but for the chain of events touched off by the simple act of a mother caring for her dying baby. Henry Whitehead had traced the river of cholera to its headwaters.

In his efforts to discredit Snow, Henry Whitehead had proved himself to be the doctor's most apt and disciplined student. The two had worked closely together for months and a lasting friendship founded on mutual respect emerged. Whitehead held Snow in such high esteem that twenty years later a portrait of his friend was still hanging on his wall, a portrait that, he said, "ever serves to remind me that in any profession the highest order of work is achieved not by fussy demand for 'something to be done,' but by patient study of the eternal laws."

On one day during that long winter of 1854, as the two men worked together to understand the "eternal laws" that governed the spread of cholera, Snow turned to Whitehead and said, "You and I may not live to see the day, and my name may be forgotten when it comes, but the time will arrive when great outbreaks of cholera will be things of the past; and it is the knowledge of the way in which the disease is propagated which will cause them to disappear."

6

THE GREAT STINK

John Snow's scientific writings cast a narrow, but bright, shaft of light on the workings of his remarkable mind. His heart, however, lies hidden in the shadows of history. He never married and left no children to preserve any letters or personal diaries. His closest correspondent was probably his uncle Charles Empson, but he too died childless and unwed, so Snow's letters to him were lost. To reconstruct the man, we have only a few frozen images offered by the remembrances of colleagues.

Of all the unrecorded moments in his life, few, if any, would have better illuminated the hidden soul of Dr. Snow than a single long conversation that took place in September 1856, as he and his uncle sailed to Calais and then rolled on through western France. Almost thirty years had passed since Empson had launched the adolescent John Snow on his remarkable climb to the heights of British medicine. Now as they passed vineyards heavy with grapes and wheat fields ripe with grain, the two men could reflect on the arc of his formidable career. If the reserved, self-contained Dr. Snow ever shared his innermost feelings, his frustration at the resistance to his ideas on cholera or the pleasure he found in his success as an anesthesiologist, he would have done so on their long journey from London to Paris.

We do know that the years since Newcastle had mellowed the strong-minded Dr. Snow. The stringent teetotaler now took an occasional glass of wine. The man who had so valued the truth that he refused to read fiction could now be seen at the opera. In 1856 the Yorkshire boy left the shores of Britain for the first time in his life and landed in a world wholly unlike Victorian England.

Upon arriving in Paris, Snow and his uncle found themselves in the midst of a construction site that seemed to have no end. In 1851, just five years earlier, Napoleon III had resolved to correct what he saw as the city's imperfections. Taking a pen to the map, he had extended, connected, and redirected major streets as if the existing buildings in the city were nothing more than sand. With a flick of his imperial wrist, he had mobilized armies of workers to raze entire city blocks and relocate landmarks that obstructed his grand vision. As Snow and Empson explored Paris, demolition and construction crews were reshaping the rue de Rivoli, the rue de Strasbourg, the rue de Rennes, and the boulevard St. Michel. Before the construction crews were finished, they would create the central markets and the Bois de Boulogne, demolish and rebuild the Paris Opera House, and create a dramatic connection between the Palace of the Tuileries and the Louvre. But Charles Empson had not invited his favorite nephew to Paris to see the sights. He had come to introduce him to an old friend. John Snow, anesthetist to the queen of England, was about to meet the emperor of France.

Empson had befriended Napoleon while the would-be emperor was living in exile in Bath. After his dramatic return to power, Napoleon III remembered his old friend and welcomed him to Paris. Empson must have believed that his nephew's ideas would interest Napoleon as he reconstructed Paris. The last cholera epidemic had killed nineteen thousand Parisians and most of the city still drank from the Seine. With 250 miles of streets, the city had only 82 miles of sewers. On the long ride to Paris, Snow had carried a copy of his monograph, *On the Mode of Communication of Cholera*.

There is no definitive proof that Snow's conversation with Napoleon influenced the reconstruction of Paris, but by the time the dust settled, the city would have two hundred miles of new sewers and two new water supplies. Each new water source required pipelines more than eighty miles long. One of them ran eighty-one miles to the Marne with some sixty tunnels and bridges and eleven miles of siphons. The other pipeline took an even longer and more tortuous course to the pure waters of Vanne.

Despite Snow's royal connections in England and France, acceptance of his discoveries continued to prove elusive. He left his monograph at the Institute of France, which was offering ten thousand francs for any discovery that offered a means to treat or prevent cholera. The French medical establishment, however, found nothing worthy in Snow's heretical ideas and never acknowledged his submission.

Two years earlier, on a chilly December day in 1854, John Snow had walked into the Soho office of John Churchill and Sons and handed them the precisely edited manuscript that would become the monograph he carried to Paris. It had tripled in length since his original version to make room for the mountain of evidence he had accumulated in the interim. His commentary brimmed with growing confidence in his theory. He drew on reports from throughout Great Britain to make his case and included detailed analyses of the role of water in all three epidemics in London including his study of the Broad Street pump. The centerpiece of this new monograph was an extended report of his research on the water supply of South London.

Through the fall of that year, even as he worked on his study of the Broad Street pump, John Snow had directed most of his energy into determining the water supply of cholera victims on the south bank of the Thames. Again and again he returned to the homes of Brixton, Clapham, and Waterloo and stood in the doorways of cholera's victims asking survivors about the source of their water. If, as was often the case, they did not know, he would fill a vial with tap water to bring back to his lab.

When all the samples had been tested and the data tabulated, the results confirmed the preliminary findings he had shared with William Farr the preceding summer. Those residents of south London who had the misfortune to purchase their water from the Southwark and Vauxhall Water Company, which drew water from the Thames in the heart of London, died from cholera at a rate four times that of their neighbors whose water came from Lambeth. More than anything else, this evidence justified the considerable expense of publishing the revision to his monograph.

Snow never received a penny for his research on cholera, and for a man who was far from wealthy, his expenses were not trivial. He had not only hired an assistant for the first time, but had also drastically reduced his clinical activities and lost several months of income. Even the cost of publishing the monograph came from his own shallow pockets.

Every time Snow had shown a relationship between water and cholera, his critics argued that some factor other than water distinguished the victims of the disease from their unafflicted neighbors. In the South London study, he defeated this objection by providing detailed data on two groups of people with only two conceivable differences between them: their water supply and their death rate from cholera. Nonetheless John Snow now understood his opposition far better than he had six years earlier. He had learned that old ideas do not yield quickly or easily. Proponents of those ideas had first ignored him and then attempted to dismiss his findings. When his data proved stubborn, they had tried to twist and distort their old theories to make room for a new set of facts. Snow had seen all this and was fully prepared to see it happen again. He could never have prepared himself, however, for the grave injustice that lay ahead.

The first two official reports on the 1854 epidemic followed predictable lines. The first came from the medical council to the General Board of Health (GBH). As president of the GBH, Benjamin Hall had stacked the council with ardent sanitarians, including Neil Arnott, who had accompanied Chadwick on his tour of the slums of Glasgow and was now physician to the queen; William Baly, who still served as physician to

Millbank Prison; William Farr, from the Registrar General's Office; and John Simon, Officer of Health in the City of London. John Sutherland, one of Chadwick's closest advisers, wrote the report. Chadwick was gone from the GBH, but his political ghost remained firmly in control.

Writing on behalf of the medical council, Sutherland described their efforts to evaluate the sources of odor and the quality of ventilation in neighborhoods where cholera struck hardest. The report opened the door to the possibility that water contributed to the outbreak, but only a crack. Contaminated water was one of many factors that predisposed people to cholera. They offered the following to local health officials:

PRECAUTIONS AGAINST CHOLERA.

1. Apply for medicine immediately to stop looseness of the bowels, or it may bring on cholera.
2. Do not take any strong opening medicine without medical advice.
3. Beware of drink, for excess in beer, wine, or spirits is likely to be followed by cholera.
4. Drink no water that has not been boiled; and avoid that which is not quite clear and well tasted.
5. Avoid eating meat that is tainted, decayed or unripe fruit, and stale fish or raw vegetables. Cooked vegetables, or ripe and cooked fruit, in moderation, are a necessary part of diet at all times.
6. Avoid fasting too long; be moderate at meals.
7. Avoid great fatigue, and getting heated and then chilled.
8. Avoid getting wet, or remaining in wet clothes.
9. Keep yourself clean, and your body and feet as dry and as warm as your means and occupation will permit.
10. Keep your rooms well cleaned and lime-washed; remove all dirt and impurities immediatately.
11. Keep your windows open as much as possible to admit fresh air; and, if necessary, use chloride of lime or zinc to remove any offensive smells.
12. If there are any dust or dirt heaps, foul drains, bad smells, or other nuisances in the house or neighborhood, make complaint without delay to the local authorities having legal power to remove them, or if there be no such authorities, or if you do not know who they are, complain to the Board of Guardians.

Item 4, though a welcome addition to the list, is buried in a manure pile of sanitarian misperception.

The report from Sutherland, however, was only a placeholder for a much longer report from the General Board of Health, which it issued in July 1855. Its appendices alone stretched over 320 pages. An analysis of the meteorology of London during the epidemic accounted for more than a third of those pages and set the tone for the report. Temperature, air pressure, wind speed, wind direction, atmospheric electricity, cloud cover, rainfall, and ozone concentrations were reviewed in excruciating detail.

While avoiding reference to John Snow, the report gave special attention to what it referred to as the outbreak at Golden Square. An odd title as only one resident of Golden Square had died as compared to 108 people who had lived or worked on the five blocks of Broad Street. Even the name suggested a preemptive effort to dismiss Snow's theory. Instead the board members who had combed through the area of the outbreak catalogued the sources of odor and organic matter in the district. They suggested, for example, that the blood and entrails from the seven sheep and five oxen killed each day at the busy slaughterhouse on Marshall Street had helped launch the disaster. They worried about the effluvia from the twenty-seven dogs that lived in a single room at 38 Silver Street. They bemoaned the number of apartments without rear windows where the lack of cross-ventilation allowed miasmas to accumulate with such deadly results.

In keeping with Sutherland's report, the board did not ignore the role of water in the epidemic. At their behest London's finest microscopist, Arthur Hassall, examined almost one hundred water samples from throughout London, including the water of the Broad Street pump, under his compound lens. With the exception of water from a few natural springs and the city's deepest wells, every sample teemed with microorganisms. The Broad Street pump appeared to have far less organic matter and far fewer microorganisms than most of the city's water sources including comparable wells. Based on this finding, the authors of the report dismissed Snow's indictment of the Broad Street pump.

Snow, while disappointed, could not have been entirely surprised by the reports from Sutherland and the GBH. He understood the political strength and obstinate convictions of the sanitarians better than anyone in England. He might even have taken solace in their admission that drinking water might play a role in the spread of cholera. He did not know it, but the worst was yet to come.

In May 1856 Dr. John Simon presented Benjamin Hall with an addendum to the General Board of Health's report on the cholera epidemic of 1854. It was an examination of the water supply of South London. Simon had used the authority and resources of the GBH to obtain precise information on the number of customers served by each company in each subdistrict south of the Thames. Snow had no access to these critical numbers. Without them he could not calculate cholera mortality rates with certainty. Instead he could only estimate them and had published values he knew to be flawed. Simon, on the other hand, could duplicate Snow's study with more accurate calculations of cholera mortality rates. Nonetheless Simon's findings were similar to Snow's. He determined that users of the S&V water died from cholera at three times the rate of those who consumed the far cleaner water of Lambeth.

Simon concluded that this evidence was so strong that it would be "accepted as the final solution of any existing uncertainty as to the dangerousness of putrefiable drinking-water during visitations of epidemic cholera." Reading the report one might have assumed that Simon was preparing to concede the validity of Snow's theory. In his next breath, however, Simon shattered any such illusion. The results, suggested Simon, demonstrated that rotting organic matter in water could pose the same threat as rotting organic matter on land. Fermentation remained the common process that "brews poison." Simon had claimed the territory first discovered by Snow and then firmly planted the flag of the sanitarians on its soil.

Simon and his colleague were intent on maintaining the theoretical underpinnings of sanitary reform. Snow posed a threat to their science and had criticized their actions. Simon responded by denying not only Snow's theory, but his very existence as well. Even the discussion of the Broad Street outbreak fails to mention Snow.

One can forgive Simon's dogged adherence to sanitarian thinking. One might even forgive his callous strategy of refusing to recognize Snow, but this was not his most grievous sin. In reproducing Snow's study of South London water, Simon presents it as if it were entirely his own conception. Snow does not even merit a footnote.

If the steady stream of minor slights and grave injustices angered or even discouraged the long-suffering Dr. Snow, he had never revealed his feelings in any public forum. But this sort of gross abuse must have wounded him. Simon had stolen Snow's ideas, presented them as his own, and twisted the results to support the agenda of the sanitarians.

Still Snow held his tongue. A colleague later recalled that "he possessed a temper which nothing could provoke." Snow simply used the data that Simon had provided to update his own work. He published a paper in the *Journal of Public Health and Sanitary Review* that reanalyzed Simon's data and showed that Simon, despite having more accurate data, had made an error in calculation. S&V customers had a cholera rate six times that of Lambeth customers, not three times, as Simon had concluded.

Snow's one contemporary biographer, Dr. Benjamin Richardson, once wrote, "The experiences of life, instead of entwining around him the vices of the world, had weaned him from the world." Indeed Snow's resilience in the face of the ignorant and relentless onslaught of his detractors seems superhuman.

Richardson had come to know Snow through the meetings of the London Epidemiological Society, of which they were both members. He developed not just an admiration for Snow's scientific ideas, but also a genuine affection for the man. When the moment came, Richardson led the charge in Snow's defense.

In October 1856, more than a year after the release of Simon's report, Benjamin Richardson sat listening to a speech at the meeting of the British Medical Association in Birmingham. At the podium Dr. T. Bell Salter, a prominent London physician, was presenting a talk on the state of knowledge concerning epidemics. Salter went through the details of Simon's study, giving full credit for the discovery of the link between water and cholera to Simon and the General Board of Health. Richardson listened attentively, waiting to hear the name of his friend and colleague. As Salter concluded, Richardson was horrified. Salter had perpetuated Simon's crime by not even mentioning Dr. Snow.

Richardson rose and stepped to the podium. He chose his words carefully. "I move," he began, "that the cordial thanks of the meeting be given to Dr. Bell Salter for his learned address." Such a motion was routine, but Richardson would not stop there.

John Snow had just returned from his trip to Paris and was not at the meeting. Even if he had been present, he would probably not have risen in simple and direct self-defense. It was not his style. His friend, however, had watched him endure the indignities of the sanitarians long enough. Benjamin Richardson continued his motion:

> [T]he author of the paper has, I think, made an accidental omission in speaking of the Report of the Board of Health on the influence of the Southwark and Vauxhall water supply on cholera, in the last epidemic of that disease in London. It is well known to all who are acquainted with the subject in its fullness, that the discovery of the connection between water supply and cholera in no way belongs to the Board of Health, but exclusively to one of our own associates—Dr. John Snow."

As he spoke a rumble ran through the audience. With the mention of Snow's name, the members of the British Medical Association erupted, clamoring, "Hear, hear!" They, too, had heard enough.

Richardson went on, "The Board of Health has, indeed, up to a late period, ignored, to a great extent, this important question and it was not until Dr. Snow had, with unwearied industry, with that true genius for observation which so characterizes his labours, and at great pecuniary cost, placed the question beyond dispute."

Then he cut to the heart of Simon's deceit. "The report was nothing more than a corroboration of Dr. Snow's important and original views; and I think it by no means fair that, while the views of other men were referred to, the claims of our associate were entirely overlooked." The audience again roared its agreement, "Hear, hear!"

Having said his piece, Richardson apologized for his break from decorum. "I think it is but honest to put the meeting fully in possession of these facts and regret that I should have been obliged to digress from the simple business of proposing the resolution placed in my hands."

As the full meaning of Richardson's rebuke of Simon and the GBH rippled through the audience, a figure rose from its midst. Dr. Edwin Lankester had sat on the Vestry of St. James Parish when Snow had pleaded with them to close the Broad Street pump. At the time, even though he agreed to the closure of the pump, Lankester had found Snow's arguments unconvincing. But this time he rose in Snow's defense. "I second the motion," he said.

A few moments later, a third figure rose from his chair. It was another member of the British medical elite, Dr. William Budd. Budd had written his own monograph on cholera in which he suggested that microscopic fungus in water could spread the disease if they were inhaled. He had published his ideas shortly after Snow's first monograph. Budd had been gracious with respect to Snow's primacy then and remained so. "Certainly," he offered, "in regard to the spread of cholera by water, [the GBH] has only declared an opinion when the question had been satisfactorily proved by others and I regret exceedingly to see that Dr. Snow's great labours had been so completely unrecognized."

At that point Richardson put the motion to a vote. When he asked for supporters, a chorus of affirmation went up. When he asked for dissenters, the audience was silent. Snow's persistence and quiet wisdom had won the day. In the community of medicine, the one Snow cared about most, the tide had begun to turn in his direction.

The tenth day of June 1858 found John Snow in his office on Sackville Street, assembling another monograph. This was a summation of his work on anesthesia. Cholera was, for the moment, gone from England and the debate about its cause had receded with the disease. Snow had not been idle.

In the years since the cholera epidemic, he had continued to move forward with a broad array of research projects. He had recently proposed a new theory on the nature of cancer, offering the prescient notion that it was derived from tissues within the body due to the influence of factors in the environment and implicating nutrition as a primary cause. A year earlier he had once again administered chloroform to Queen Victoria, this time for the birth of Princess Beatrice. Just a day earlier, he had met with a group of colleagues to plan a study on the nature and cause of the sounds of the heart as revealed by recent improvements in the stethoscope. In a moment when Snow seemed on the verge of extending the reach of his brilliance into new fields of medicine, tragedy struck.

He had just dipped his pen in ink and scratched the word *exit* when he slumped in his chair. A stroke held him in its terrifying grip, and he was unable to move. The stroke was relatively mild. By the next day, he had recovered enough to continue work on his monograph, and it seemed, for a moment, that his remarkable career might continue.

Two days later, however, his housekeeper entered his office to find him on the floor, unable to move his left arm and leg. His face was contorted, with his mouth drawn to one side. This time the paralysis persisted. Snow remained in bed for several days, lucid, but unable to move half of his body. On June 16, the life and career of John Snow ended, perhaps as the result of a third stroke. He was forty-five years old.

Even in his death, his detractors seemed determined to diminish him. The *Lancet* offered the following, two-sentence obituary:

> Dr. John Snow: This well-known physician died at noon on the 16th instant at his house on Sackville Street from an attack of apoplexy. His researches on chloroform and other anaesthetics were appreciated by the profession.

Fainter praise rarely has been given.

If he had lived another twenty years, we might well find his name next to Darwin and Pasteur as one of the greatest scientists of his age. But this cruel and abrupt end to his life, together with the disappearance of any records of his personal life, has helped to relegate him to the second tier of scientific history.

He did, however, leave us with an idea. The idea that organisms in drinking water could cause disease would continue to be controversial for more than thirty years after Snow's death. Some never accepted his ideas. In 1894 William Chadwick went to his grave, resolute in his sanitarian beliefs.

The problems with London's water supply as identified by John Snow would not begin to go away until the wholesale dumping of sewage into the Thames stopped. All the evidence Snow produced did little to convince members of Parliament to take the costly steps needed to improve the quality of the water in the river. Then, in the months after Snow's death, nature forced the issue.

The summer of 1858 was long, hot, and dry. The end of July found the parched Thames struggling to reach the sea. The shallow remnants of the river ebbed and flowed with the tide, offering no escape to the steady influx of sewage. The Thames became a vast, sloshing cesspool. As the August sun baked the river, it gave off a stench so overpowering that it became known as the Great Stink, perhaps the only time in history that a smell earned itself a proper name.

That summer members of Parliament retreated from their usual chamber to remote meeting rooms as far as possible from the river's bank. Breathing through scented handkerchiefs, they voted to approve the construction of a vast set of interceptor pipes to carry London's sewage eighteen miles downstream, far from the intakes of the water companies. Within a year construction began on what was then the largest engineering project in British history. Once it was completed, cholera left England, never to return. Still, the debate over cholera's cause and its link to drinking water would not subside until the deadly bacteria had been found.

Robert Koch *Vibrio cholerae*

Cryptosporidium emerging from
oocysts in the small intestine

PART II

Thirsty Cities and Dirty Water

"... the good feeling was as universal and exuberant as
though the Lake Tunnel had flooded the city with champagne
and oysters, instead of pure water without little fish."

THE *CHICAGO TRIBUNE,* ON MARCH 26, 1867, UPON
THE INAUGURATION OF A NEW WATER SUPPLY.

THE RACE TO CHOLERA

The footsteps of three men echoed up the stairwell of the Hôpi-tal Larboisière. They climbed steadily, past the wards of one of Paris's newest and finest medical facilities. One of them, a short man in a black suit with a crisp bow tie and a trim beard, led the group as they entered the attic. The men formed a select team, sent by the French government to find a killer. Their leader that day, the youngest of the three and the group's only nonphysician, was a chem-ist by the name of Louis Pasteur.

At forty-three, Pasteur had just completed a series of studies for France's food and beverage industry that had laid the foundation for an entirely new field of science that would come to be known as micro-biology. In that attic Pasteur had been conducting his latest study. He led his colleagues to a ventilation shaft that rose from the wards below. Peering through his pince-nez spectacles, he bent down and attached a glass tube to an opening in the shaft. The tube ran through a mixture of refrigerants to a small hand-powered fan. Cooling the air would cause water vapor to condense, which in turn would cause air-borne particles and microbes to collect along the sides of the tube. As he cranked, the fan pulled air through the tube from the ward below where, one after another, patients were dying of cholera.

It was October 1865 and the dreaded disease had once again bro-ken loose on the streets of Paris. The minister of health had sent the finest team available in search of cholera's cause. The stakes were high. The epidemic was slaughtering some two hundred Parisians with every passing day. As they tested the air, the deadly waters of the Seine ran through the plumbing of Paris. A new water supply was almost a decade away.

When he received the minister's call, Pasteur was immersed in a study of diseases in silkworms. He had never studied human disease before. But Pasteur did not shy from the challenge. In addition to the samples from the ventilation shaft, he sealed flasks of air taken in spots throughout the hospital. The three scientists also collected blood from several cholera patients before returning to their laboratory. They even gathered dust from the floors. Somewhere in these samples they hoped they might find the cause of the deadly outbreak. The simple fact that they were analyzing air and blood in search of cholera's spark, however, testifies to the level of heresy present in John Snow's original work on cholera, published sixteen years earlier. Three of the finest scientists in France were looking for the organism in the two places Snow had insisted they would not find it.

They of course failed to find cholera in the air that rose from the cholera ward, but one can forgive Pasteur's obsession with the air as a vector for transmitting disease. Among the most remarkable findings of his earlier research was the presence of bacteria and mold in the air wherever he looked, from the brine pits of Arc-et-Senans to the glaciers of Mont Blanc. Pasteur, however, had never studied microbes in water. The culprit responsible for some nine thousand deaths in France would leave the country with the secret of its identity intact.

In the eighteen years that passed before Pasteur again set out to find the cause of cholera, he led his growing band of apprentices through what he called "the unexplored country," a land that he had in large measure discovered. Soon others began to probe this uncharted territory. In time Pasteur found himself on a collision course with the other founder of modern microbiology, a young German physician by the name of Robert Koch. The conflict began as a war of words but became a race to find the cause of cholera, a contest that would leave a promising young scientist dead in the Egyptian desert before ending in a water tank north of Calcutta.

In the fall of 1865, as Pasteur struggled to unravel the cause of cholera, a young Robert Koch sat down to eat a plate of butter. In the city of Göttingen, in the north of Germany, cholera was far from his thoughts. He pulled himself up to the table and began to consume this most unusual meal. Stuffing spoonful after spoonful of butter into his mouth, he did not stop until he had consumed half a pound. At the time the twenty-two-year-old medical student seemed like an unlikely competitor to the great French scientist, but this bizarre meal was the beginning of Robert Koch's research career. He was studying the metabolism of dietary fat. As soon as he had finished dining on butter, he began collecting and analyzing his own urine.

By the time he finished medical school Koch had already written two research papers. In addition to his work on diet and urine chemistry, he completed an exquisitely detailed dissection of the nerves of the uterus, which he wrote up, illustrated with meticulous drawings, and emblazoned with his personal motto, "Nunquam otiosus" (never idle). But Koch had not intended to become a researcher and would not write another scientific paper for more than ten years.

He began those ten years bouncing from one small town to the next with his wife, Emmy, and their young daughter, Trudy, determined to support them on his income as a general practitioner. He finally found some stability as the district medical officer in Wollstein, a city in the Polish-speaking region of eastern Germany. He and his family moved into a solid two-story home with a horse barn and a large garden in the back.

True to his motto, Koch was indeed never idle. He played the zither, bowled, socialized, and climbed the local mountains. He even established himself as a skilled amateur archaeologist. He accumulated a veritable zoo of pets and livestock ranging from pigeons and chickens to bees, rabbits, and even a pair of monkeys. Above all he was a superb physician. His patients loved the hardworking young doctor and flocked to the clinic he established in one half of a large, well-lit room that stretched across the south side of the house. In the other half of the room, hidden by a curtain, his future began to take shape.

The life of a country doctor could not hold the ferocious intellect of Dr. Koch. As his financial situation improved and he was able to save money, his first purchase was not the carriage that he sorely needed for house calls in rural Germany, but a fine microscope. He placed it in the room behind the curtain, which was coming to look more and more like a laboratory.

Far from the intellectual ferment of the late nineteenth century, Koch embarked on a solitary quest for understanding that began with the revolutionary ideas of Pasteur and led to a rigorous framework for identifying the cause of specific diseases. Koch took the world around him and placed it under his microscope. Every free moment found him peering into the microcosm that lay hidden in everything from pond water to drops of his patients' blood to the organs of sick sheep. Through painstaking trial and error, Koch learned which dyes to add to his slides to create stark outlines around transparent microbes. The barn behind the house provided a steady supply of wild mice as he began to experiment with the transmission of infections. He had already developed a variety of methods for growing bacteria when late in 1875 a message arrived at his door. He was wanted at the police station.

When he arrived the anxious officer escorted him to the object of his concern, the bloody hide of a sheep. A local farmer had tried to salvage the hide from an animal that had collapsed in a field with blood pouring from its mouth and nostrils. The farmer, the policemen, and Koch knew all too well what had happened. The animal had died of anthrax.

The three years Koch had spent building his laboratory in Wollstein and developing methods for studying bacteria were preparation for this moment. Throughout the nineteenth century, anthrax infections routinely struck at the farms of Europe with devastating consequences. Other scientists had seen the bacteria responsible, but no one understood how the disease could strike suddenly at animals that had had no contact with victims of the disease.

To study anthrax Koch would first need to transmit the disease to a new animal. This would assure him a supply of viable bacteria while he tried to culture them in the laboratory. On December 23 he selected a rabbit from among the animals that his wife, Emmy, maintained for him in the cages that filled their garden and horse barn. He drew a bit of the blood from the hide into a glass syringe, held the squirming rabbit with one hand, and injected a large vein in its left ear. By Christmas Eve the rabbit was dead.

One month later Koch pulled back the curtain and entered his lab carrying a large jar of eyeballs. He set to work draining the fluid from each one and transferring it into glass culture tubes. He had discovered that the fluid provided a perfect culture medium for growing anthrax and had just returned from one of his routine trips to the local slaughterhouse to collect the eyes from cows, pigs, and sheep. He had even designed and constructed culture tubes for the specific purpose of growing the deadly bacteria. In one corner of his laboratory, a kerosene lamp warmed a bed of sand. Koch had built it after realizing the bacteria grew best at temperatures close to ninety degrees. All day long he monitored his cultures and adjusted the lamp.

Within a month in his makeshift laboratory in a world without electricity he had worked out the life cycle of the bacteria including the crucial observation that anthrax could form spores. The spores proved to be both highly infectious and durable enough to remain viable for years in a harsh environment. This explained why simple exposure to soil long after an area was free from anthrax could cause disease. When he brought this landmark work to the world's attention, the rigor of the methods and the level of innovation coming from a man working in almost complete isolation stunned the scientific establishment and launched Koch's career.

Five years later Koch joined a select group of the most prominent medical researchers in the world as they crowded into a laboratory at King's College in London. The laboratory belonged to Joseph Lister. Lister had introduced sterile techniques to surgery, an innovation that, together with anesthesia, marked the beginning of modern surgery. A tall man with a handsome face framed by great bushy muttonchop sideburns, Lister commanded the attention of the eminent scientists he had invited to his laboratory, but he was not the man they had come to see. They had come to see Robert Koch.

Before the attentive crowd, Koch demonstrated a technique he had developed for growing colonies of bacteria on plates of solid agar. A London physician stood to one side, translating Koch's German into English. Koch selected individual colonies and demonstrated how one could obtain pure cultures of a specific bacterium with relative ease. Pure cultures were essential to meaningful research. The audience, all of whom had struggled with complex and often unsuccessful methods for selecting a single bacterium from a liquid broth, watched intently. In liquid broth the mixture of bacteria formed an almost inseparable soup. On Koch's agar individual bacteria grew into homogeneous, well-defined colonies.

The audience of the world's finest microbiologists immediately grasped the value of Koch's work. When he finished a man with a marked limp made his way forward to shake his hand. *"C'est un grand progrès, monsieur!"* he said, and extended his hand. Koch shook it and for the first time in his life looked into the eyes of Louis Pasteur.

Both men were in London to attend the International Medical Congress. The universe of medicine was in the midst of an explosion of new ideas and techniques and all its luminaries were there. At age fifty-nine Pasteur was among the brightest stars in that universe. Koch at thirty-eight was ascendant. Just one year earlier, he had joined the Imperial Health Office in Berlin as the director of a group of microbiologists. He and Pasteur, however, were cut from different cloth. Koch had founded his accomplishments on relentless precision, a technique so imbued with care that it had allowed him to work safely with an agent as deadly as anthrax in a makeshift nineteenth-century laboratory. He brought to microbiology a rigor never before seen and reinvented the tools and techniques essential for the science. Pasteur was a careful scientist, but his brilliance derived from a series of bold cognitive leaps. In many ways the two men personified their countries and, in the months that followed the London meeting, they would reenact the war that had engulfed those countries ten years earlier.

On the second to last day of the congress, Koch walked into St. James Hall in the heart of London to hear the words of Pasteur as he described, for the first time, the development of a technique to generate vaccines, one of the most important innovations in medical history. Koch struggled to hear Pasteur's unamplified words as they rose up toward the sixty-foot-high ceilings of the hall. As Koch listened Pasteur also described his first successful application of the new method. Pasteur, he discovered, had developed and tested a vaccine for anthrax.

Koch was stunned, but he could not listen to Pasteur with unbiased ears. He and both of his closest colleagues at the Imperial Health Office in Berlin had fought in the Franco-Prussian War. Although ten years had passed since the war, its wounds still festered. The antipathy between the countries remained so intense that when Lister arranged for a banquet for delegates to the congress, he had to hold two, one for the French and one for the Germans.

Nonetheless, it is hard to find anything in Pasteur's speech that Koch could take as an insult. Undeterred, Koch took offense. Perhaps he expected greater recognition from Pasteur for his accomplishments.

Perhaps he saw anthrax as somehow his disease. Whatever the reason, the battle was joined.

Just two months after his return to Berlin, Koch published a blistering nine-page attack ridiculing Pasteur's work on anthrax. It focused on what he saw as the lack of precision in Pasteur's methods. A single, one-sentence paragraph from that paper sums up Koch's position best. "Only a few of Pasteur's beliefs about anthrax are new," he wrote, "and they are false."

If Koch's intention had been to get the attention of Pasteur, he succeeded. Pasteur, the son of a soldier in Napoleon Bonaparte's army, had not fought in the war, but could never forgive the Germans for invading France and precipitating the fall of Napoleon III. He had not started the battle with Koch, but he did not shy away from it either. A bitter bilingual debate ensued as the two men took turns firing salvos at each other in the French and German medical literature. Koch's critique of Pasteur focused on weaknesses in his methods and was well founded except for one thing. The vaccine worked.

In the spring of 1882, a young French physicist by the name of Louis Thuillier drove the needle of his syringe through the hide of a sheep in the Prussian village of Packish and injected one sixth of its contents into the animal. Fifty sheep and a dozen cows bleated, mooed, and clattered across the floor as Thuillier moved from animal to animal.

Thuillier had joined Pasteur's lab just two years earlier and at twenty-six he was still its youngest scientist. His careful, precise manner in the lab had quickly made him a favorite of Pasteur who enlisted him to work on his most important projects. Pasteur had already sent his bright young star to Budapest to demonstrate the vaccine when he received a request from the Prussian government for a similar trial. Given his conflict with Koch, this was a delicate assignment. It would be conducted under a microscope of scrutiny with every detail of the experiment specified, right down to the concrete floor, an almost unheard-of feature, which would ensure that anthrax in contaminated soil could not disrupt the experiment. Pasteur had not hesitated to send Thuillier again.

As the small man with meticulously cropped hair and an explosive beard refilled his syringe with a sample of Pasteur's precious vaccine, a cluster of eminent Prussian scientists watched intently. The observers followed Thuillier from animal to animal, noting his every move.

The trial would ultimately prove the value of Pasteur's vaccine. Thuillier would so impress his handlers that he would receive the Cross of Knight of the Crown of Prussia for his work, but in the barn in Packish he worked under a cloud of uncertainty. The demonstration in Budapest had not gone smoothly, and success in Germany was far from assured as he rode back to Berlin, three hours by train and carriage, with his satchel of used syringes and empty vials of vaccine.

After arriving in Berlin, Thuillier could only wait for the vaccine to generate immunity. Thuillier, like his mentor, was not one for idle moments. Determined to make use of his time and driven by his intense curiosity, he made the trip across the city to an unremarkable brick building just across from the Charité, Berlin's largest and most famous hospital. There on the third floor, in a small, crowded laboratory, he found Robert Koch.

In the same room, in a small dish, Thuillier found Koch's newest and most famous conquest, the bacterium responsible for tuberculosis, a discovery that would eventually earn Koch the Nobel Prize. Thullier had received a prepublication copy of Koch's paper on tuberculosis and could not resist a visit. This journey into the enemy camp must have been awkward at best. In writing about it, Thuillier seems to have been torn between his prejudices as a member of Pasteur's camp and his natural predisposition to objectivity as a physicist. Despite a devotion to his mentor, an adulation he expressed in every letter he sent to France, Thuillier came away from Koch's laboratory duly impressed. The most biting critique he could offer was a note to Pasteur describing Koch as "a bit of a rustic." Koch kept no record of the meeting, but future events would suggest that he had been impressed by the serious young scientist as well. That future would involve a meeting of the two men in a place far from Berlin under far different circumstances.

On August 16, 1883, Robert Koch climbed onto a train bound for Italy as porters loaded it with nine large wooden crates. Koch and his colleagues at the Imperial Health Office had spent the past week filling the crates with everything from microscopes and cover glass to sheep's blood for culturing bacteria and wire mesh for covering mouse jars. They had packed everything they imagined they would need to track down a pathogen that none of them had ever studied in a land none of them had ever seen. Koch and his team would accompany the crates to Egypt, where they would begin the hunt for cholera.

Cholera had slithered out of its nest in India once again and, with the opening of the Suez Canal in 1869, Europe no longer had the mountains of Asia and the long sailing trip around the horn of Africa to protect her. The French, upon learning that cholera had erupted in Egypt, had rushed to send a team to investigate. French ports were threatened. France had colonial roots in Egypt including their leadership role in construction of the canal, and in the wake of their humiliating defeat by Germany in the Franco-Prussian War, they wanted to find cholera before the Germans. Not to be outdone, the German government quickly assembled its own team, with Koch as its leader.

Koch now had an opportunity to move beyond his war of words with Pasteur and prove, once and for all, his superiority in the laboratory. Koch and his team rushed to assemble and pack the components of a state-of-the-art microbiological laboratory and, just a week after learning about the French expedition, he was on his way to Alexandria with three of the most skilled members of his research team. No amount of preparation could erase the fact that they had already given the French team a two-week head start.

The train wound through the Alps, down to Verona, and on to the port of Brindisi where they boarded the steamer *Mongolia*. The heat grew with each day as Koch's team sailed south to Port Said at the mouth of the Suez Canal. The Mediterranean was calm, but Koch spent much of the journey doubled over with seasickness. For this man who was thirty-six before he even saw the sea, the solid ground of Egypt was a welcome sight.

After a day of quarantine in Port Said, the team boarded an Egyptian steamer for the trip to Alexandria. On August 24, eight days after they left Berlin, carts laden with wooden boxes and luggage rumbled through the busy streets of Alexandria from the Great Harbor to the Greek Hospital. With the completion of the canal, the city's population had quintupled in just forty years and its narrow streets teemed with people and animals. As soon as the crates arrived in the sunlit room that would serve as their laboratory, they set to work unpacking their precious equipment and supplies. As the lab took shape, the room buzzed with the anticipation of the hunt.

That evening, as Koch's team pried open the wooden crates, a thirty-two-year-old Sudanese man arrived at the hospital, anxious, weak, and unable to speak. He had been vomiting and his stools were loose and bloody. His skin was cold and his pulse faint. He was sent to the cholera ward, but the care he received there did nothing to improve his condition and, at eleven o'clock the next morning, he died the graceless blue death. Chance had just made him the subject of Koch's first study.

Just after four o'clock on his second day in Alexandria, Robert Koch sliced open the unfortunate man's abdomen. He carefully lifted the colon from the edge of the abdominal cavity and ran a scalpel across the thin bands of muscle that surrounded it. As the contents spilled out, Koch expected to find the gut filled with the so-called rice water that typified cholera. He had planned to culture this liquid and isolate the bacteria responsible. So strong was his expectation that he had convinced himself he had seen this as a young doctor fresh from medical school during autopsies of cholera victims in the Hamburg hospital. He was mistaken.

The gut was filled with all manner of bacteria and they seemed to be in the process of devouring its lining. Koch saw little hope of isolating anything from the purulent fluid. He then noticed that the intact portions of that lining were covered with angry red patches.

He rushed back to the lab so that he could examine the red patches before the sun, the only source of light adequate for microscopy, dropped below the western desert. He placed the sample in a microtome, cut a razor-thin slice, and put it on a slide. At his microscope Koch adjusted the mirror, carefully bending the fading rays of sun until they flooded the slide with light. As he squinted into the lens, he saw hundreds of rodlike bacteria burrowing into the gut. When he switched to a higher magnification, he noticed that the rod was bent just slightly, like a comma.

For Koch seeing the bacteria was far from conclusive. By his own rigorous standards, standards that would become a central tenet of microbiology known as Koch's postulates, he would need to find the bacteria in every case of cholera, isolate it in culture, and produce cholera in another animal with the isolated organism if he wanted to demonstrate that it was the cause of cholera. The German team had a long way to go, but in their first day in Alexandria they had already made an important start.

The German team, however, was not alone. The following morning, just after eight o'clock, a team of French scientists at the European Hospital sawed through the ribcage of Luisa Bigatti. The thirty-nine-year-old Italian woman had died of cholera the night before. They pulled back her ribs and began to examine her heart and lungs. They had already probed the length of her digestive tract, stopping as they went to take samples for later examination under the microscope. The process had become routine. After ten days in Alexandria, Luisa Bigatti was their thirteenth cadaver.

Sixty years old and debilitated by a stroke that had left him partially paralyzed fifteen years earlier, Pasteur had chosen to forgo the arduous trip to Egypt. Instead he had sent two of his most trusted assistants, Emile Roux and Louis Thuillier. A tall, thoughtful physician with a sparkle in his eye and a passion for research, Roux had played a key role in Pasteur's work developing vaccines for anthrax and rabies.

Thuillier, the youngest member of the team, had been reluctant to go to Egypt. A growing romantic relationship had made him suddenly reluctant to travel so far from France. Pasteur saw the quiet, young physicist as a key member of the team and encouraged Thuillier to go. In the end, out of loyalty to his mentor, he relented.

In addition to Roux and Thuillier, Pasteur had sent a senior physician to lead the team and a skilled veterinary surgeon to manage their animal studies. His battle with Koch would be carried out by proxies, but Pasteur had given them a detailed plan of attack.

Like the Germans the French team carried out an exhaustive research program that began, but did not end, with the examination of cholera victims. They collected and examined vomit and stool samples. They placed hundreds of tissue samples under their microscopes. They scoured the bustling markets of Alexandria for test animals. They tried to induce cholera in everything from chickens, quail, and turkeys to pigs, dogs, and monkeys by injecting intestinal contents, blood, stool and even bits of the intestines themselves under the skin, into the blood, and into the gut of the hapless animals. They fed the assembled menagerie bits of the nerves, livers, spleens, kidneys, and lungs of cholera victims. Nothing seemed to work. On one occasion they suspected they had induced cholera in a chicken, but it is more likely they managed to induce some other sort of infection.

The French team had noticed the same bacteria that had caught the eye of Koch, but were more intrigued by the "small bodies" they found in the blood of cholera victims. They studied these small bodies in great detail and attempted to transmit the disease by isolating them. They even believed at one point they had cultured them. It now seems likely they were looking at nothing more remarkable than tiny, floating blood clots.

One early September evening, as the moon rose over the Nile delta, four bearded men on donkeys raced across the desert toward the sea. After a day spent in stiff European clothes that suited their status, but not the climate, the rush of cool air offered intoxicating relief from the relentless heat of the day. Robert Koch, the smallest of the four men, leaned forward, scanning the horizon as bits of sand and dust bounced off his thick glasses. Several Arab guides in flowing robes trailed behind. They galloped on into the fading light until they reached the end of the desert. There the strange caravan pulled to a halt on jagged cliffs that dropped to the waves below.

As the sky grew dark and exploded with stars, the German team sat on the cliffs above the Mediterranean around a fire prepared by their guides, eating supper and reflecting on their progress. Like the French team, they had been unable to reproduce cholera in laboratory animals, but they were far better equipped than the French to culture bacteria. They also had more experience than anyone in the world with the techniques Koch had pioneered using solid agar and the unique culture dish developed by a man named Petri who worked in his lab. They had even come prepared for the relentless desert sun. In Europe their labs had required heated incubators to grow bacteria but under the blistering Egyptian sun, the problem changed. The Germans found a supply of ice and placed it in a small ice box they had built especially for the trip in order to keep the agar in their culture media from melting.

The four Germans talked beneath the desert moon as the firelight played across their faces. Their words swung between the ripening promise of success and the looming possibility of failure. On one hand, they sensed they had found the cause of cholera. The small, comma-shaped bacteria inhabited the gut of each of the nine cholera victims they had dissected, but none of the cadavers who had died from other causes. Even more enticing, they seemed close to growing it in culture, a step that would free them from their grisly dependence on corpses. On the other hand, their research faced an almost insurmountable problem. That night, as their fire burned down, cholera was disappearing from Alexandria.

The outbreak had peaked just two days after the French team arrived, when fifty people died. On August 24, the day the Germans arrived, cholera killed twenty-two people. By September 3, they had conducted nine autopsies, but they did not get another until September 29, and that would be their last. With no material left to study, the frustrated scientists busied themselves investigating cholera in other cities. In Alexandria they could only turn their attention to other diseases.

Despite the fact that the French and German teams were working just a few blocks from each other, they had had no interaction. Even without the bitter antipathy between Koch and Pasteur that permeated the two groups, the teams might have found reasons to dislike each other. Edmond Nocard, the French veterinarian, had fought in the Franco-Prussian War, and Roux had lost two brothers in the conflict. Koch and Georg Gaffky had both served as physicians in Bismarck's army, as had Friedrich Loeffler, their colleague in Berlin. In every sense, these two groups were quietly at war. Then tragedy struck.

In the weeks without cholera, Thuillier had chosen to study rinderpest, a disease that devastated cattle herds in Europe as well as Egypt. He and a colleague spent much of the third weekend of September at a slaughter yard, examining animals that had recently died. That Monday evening Thuillier tried to escape the heat by going for a swim in the sea. After an evening carriage ride and a late dinner with Roux, he went to bed for the last time in his life.

At three that morning, Roux woke to find Thuillier standing in his room. Thuillier managed to utter the words, "I feel very ill," before collapsing face-first on the floor. Roux, together with Straus, the other doctor on the team, carried Thuillier back to his bed. He recovered his senses and they offered him opium. He drifted to sleep, but by eight o'clock that morning, he was suffering from profuse diarrhea as violent cramps tore through his body. Roux and Straus looked on in horror as their young colleague developed the obvious symptoms of cholera. The hunter had become the prey.

A group of Italian physicians came to assist and the assembled doctors agreed on a treatment regimen of "strong frictions," ether injected under the skin and a drink of iced champagne. Despite their efforts (or perhaps hastened by their efforts), the twenty-seven-year-old scientist died that evening.

The source of Thuillier's cholera is a mystery. He had had no contact with a cholera victim during the two weeks before his death, a period far longer than the incubation period for the disease. Neither could the slaughter yards have been a source for his infection as cholera does not affect livestock. Perhaps the answer lies in the safety precautions issued by Pasteur. He had wisely insisted that the team drink nothing but boiled water. He had also urged that they use boiled water to wash any fruits and vegetables. That washing would have been insufficient if Thuillier had chosen to eat unpeeled produce from a region in which the disease was active. Whatever the source of his cholera, the death of Thuillier melted the ice that had existed between the two teams.

Despite the acrimonious debate between Pasteur and Koch, the two men held a begrudging admiration for each other. The death served as a poignant reminder that their ultimate enemy was not each other, but the blind, heartless pathogen they both sought to defeat. When word of the disaster reached them at the Greek Hospital, Koch and his team rushed to offer their support to the devastated French team. In halting French Koch recalled Thuillier with admiration and kindness that moved even Roux. When they returned for the funeral, they brought two laurel wreaths, which they fastened to the coffin, a token that Koch described as "just suitable for one who deserves such glory." Then Koch himself helped carry Thuillier's casket.

The death took the heart out of the French expedition. News of it devastated Pasteur who had taken the promising young scientist under his wing and convinced the reluctant Thuillier to go on the trip. "I felt such esteem and affection for him!" he wrote. "I would have gladly asked for him the Legion of Honor! His death was glorious, heroic. Let this be our consolation." Soon after the tragedy, the team returned to France with little to show for their efforts.

Koch's team had no intention of heading home empty-handed. If the beast had fled, Koch would follow it back to its lair. He sent a request to the German government asking for support for a trip to a place where the cholera epidemic never ended. By November he was steaming across the Indian Ocean, bound for Calcutta.

Until he could culture the pathogen, Koch existed in the ethical netherworld that required a steady supply of victims to continue his research. From the moment of his arrival at the Medical College of Calcutta, a grand colonial structure in the heart of the city, it was clear that cholera would provide. He and his two colleagues had had just one day to set up their equipment when the first cholera victim arrived, an anonymous crewman from one of the many ships that plied the busy harbor, identified only as a European sailor.

During his time in Alexandria, Koch had noticed that the shorter the time between death and autopsy, the more the comma bacillus dominated the contents of the gut. If he waited too long, the gut was overrun with every manner of bacteria, and culture was not possible. When he opened the abdomen of the unfortunate sailor, he realized the autopsy had not come soon enough.

The next day another victim arrived in his autopsy room. Again it was too late. Then on the afternoon of their fourth day in Calcutta, a hearse arrived outside the laboratory building with a young Indian man who had died just two hours earlier at a nearby hospital. Koch and his team rushed to the autopsy room and moments later he found what he had been looking for. As quickly as possible, they brought the sample down the hall to the lab where brilliant afternoon sun beamed through three large windows. When he aimed the sunlight through the microscope, the comma-shaped bacteria he had seen in Alexandria filled the slide.

Between cholera's disappearance from Alexandria and Koch's arrival in Calcutta, almost three months had passed, three months in which Koch could plan for this moment when he saw cholera again. He spread some of the specimen on agar that he had enriched with beef broth. The next morning the culture plate was covered with bacteria. When he looked at them in his microscope, he found that the comma-shaped bacteria had not only grown on culture, they had replicated into spirals, curlicues, and circles that squirmed wildly on his slide. He had his pathogen.

Koch was exuberant over his finding, but over the weeks that followed, that excitement would fade as he struggled to prove that the comma-shaped bacteria not only were present in cholera's victims but also could transmit the disease. He followed the methods that had served him so well. He collected every type of animal he could find and injected one after another with massive doses of the bacteria. Each time he waited. Each time nothing happened. Cholera, as it turns out, infects only humans, the one laboratory animal unavailable to him. (Researchers would ultimately find a way to infect guinea pigs by surgically altering their intestines, but Koch never imagined that such extreme measures would be required.)

Eight weeks later a frustrated Robert Koch closed the door of the animal room in the basement of the Medical College and walked upstairs to his laboratory on the second floor. He and his colleagues had tried in every way they could imagine to infect everything from mice to monkeys with the comma bacillus, but to no avail. As he sat down in his office to assess the situation, the punkah above his head flapped slowly back and forth. The primitive fan, driven by an impoverished Indian who pulled a rope in another room, discreetly out of sight, produced a feeble current of air that did little to break the heat. With daily high temperatures rising above 100 degrees, even their culture media were starting to melt. It would soon be impossible to maintain pure cultures. Koch was certain he had found the cause of cholera, but the time he needed to generate irrefutable proof was running out. Then he received a message from the sanitary commissioner for the British Government of India.

The water tanks (more accurately imagined as small man-made ponds) of Calcutta were, in the words of a British health officer, nothing less than a "means of committing sanitary suicide." In the dry climate, these small reservoirs served as the sole repository of water for every purpose from bathing and washing clothes to drinking and cooking. The sanitary commissioner informed Koch that a cholera outbreak had killed seventeen people around one of these tanks. When Koch visited the tank, he could see how wastewater drained from the crude privies outside the walls of the surrounding houses toward the tank. He learned how the soiled clothes of a cholera victim had been washed in the tank and he took a sample of the water. Back in his office, Koch drew a map of the neighborhood. When he marked those homes where someone had died of cholera, it looked remarkably like Snow's map of the area around the Broad Street pump. A circle of death surrounded the suspect tank.

Koch, however, with the tools of microbiology, could take this one step further than Snow. When he applied those tools to examine the water from the tank, he found it swarming with the bacteria he had seen so many times before. Nonetheless, in a twist of irony, the great bacteriologist still needed epidemiology to prove his point. Koch could not fulfill his postulates in the laboratory, but the water tank provided a natural experiment. His epidemiological study of that experiment, together with his other findings, convinced enough scientists that he could claim to have identified the agent responsible for cholera.

Koch embarked on the long trip back to Germany as a conquering hero. Kaiser Wilhelm received him upon his arrival and awarded him a medal created for the occasion. He was given an audience with Bismarck, a grand banquet, and a grant of 100,000 marks in gold for his research. But despite all of the adulation and awards, Koch's discovery did not tell the world how to stop the waves of cholera epidemics. That answer would not arrive for eight years.

On August 22, 1892, a man walked into the triangular building that housed the newly constructed Institute for Infectious Disease in Berlin under the shadow of the Charité Hospital. He anxiously clutched a leather satchel, taking special care to protect it as he climbed to the second floor. In a well-lit office on the point of the triangle, he was greeted by a balding man with a trim gray beard, the founding director of the institute, Robert Koch.

The visitor opened the satchel and with exquisite care removed a culture tube and handed it to Koch. The man, a physician, had arrived on the train that morning from the city of Altona, just downstream from Hamburg. When Koch examined the contents of the culture tube, he saw, for the first time in years, the deadly crescent of cholera.

On that same day, the director of Hamburg General Hospital rushed to the office of Theodor Kraus, the chief medical officer for the city of Hamburg, to inform him that cholera had reached that city as well. In fact the epidemic had arrived almost a week before, but a stubborn partnership of denial and incompetence had kept the physicians from culturing and recognizing the organism. The hospital director expected that Kraus would immediately alert the city. To his horror Kraus dismissed the finding and would go on to use every power at his disposal to staunch the flow of information critical to the public health. Kraus belonged to a large group of physicians and public health officials who were unconvinced that Koch's bacteria could cause cholera. This point of view sat well with the economic elite of this port city who had little stomach for quarantine.

Undaunted, the hospital director took matters into his own hands. He sent Kraus an official state telegram announcing the arrival of cholera in Hamburg. This meant that a copy would go to the central government in Berlin. Federal health officials responded by putting Dr. Koch on a train to Hamburg.

For Koch, a call about cholera must have come as a relief. As he rode through the early morning, he could, for a moment, leave the turmoil of his life in Berlin behind. Two years had passed since he had announced to great international fanfare that he had found a cure for tuberculosis. In the intervening years, the man who prided himself on thoroughness and precision had been proven rash and wrong. His earlier success had bred jealous enemies and they seized on his moment of failure. His personal life had fared no better. His marriage to Emmy was falling apart, and at fifty-five he had launched an impassioned affair with an eighteen-year-old art student.

When Koch arrived in Hamburg, Kraus refused to meet him at the train station. When he went to Kraus's office, the eminent scientist sat waiting for half an hour until Kraus arrived. In Kraus's mind, there was no epidemic and no need for Koch. His denial did not stop the death toll from reaching 320 that day.

Unwavering and a bit indignant, Koch went on to tour Hamburg with members of the city's government and medical community. What he saw left him shocked. He would later write to his young mistress that as he visited Hamburg's hospitals and saw the condition of the ships and emigrants in its harbor, he felt as if he were "walking across a battle-field." In the midst of the squalor of the city's slums, Koch could not help but recall his time in Alexandria and Calcutta. "Gentlemen," he said, turning to the other members of the entourage, "I forget I am in Europe."

Indeed the outbreak would prove far worse than the one that had struck Alexandria. Over the next week almost 3,000 people would die. Before it was done, the epidemic would claim over 8,600 lives, most of them in just three weeks. Just downstream from Hamburg, in the city of Altona, the outbreak took a very different course.

Altona and Hamburg were separate cities, but Hamburg had grown to the point where it subsumed its smaller neighbor. Both cities had similar populations living under similar conditions and both drew their water from the Elbe River. The two cities had grown together in such a way that Altona could be said to differ from Hamburg in only two important aspects. First, Altona filtered its drinking water. Second, very few of its residents died of cholera.

Altona in fact had filtered its water since 1859. Officials in Hamburg had first announced plans to filter its water in 1860, but had spent the next thirty-three years wrangling over the design of the filters, the cost of the system, and above all who should pay for it. Hamburg maintained a titular government, but commerce was king and business interests seemed to have the final say in all decisions. The head of the city's property owners association offered an assessment of the city's water in 1885, in which he concluded that "Hamburg must be the cleanest city in the world with its water supply and sewage system." The city had installed a system of sanitary sewers in 1850, after close consultation with London's sanitarian Edwin Chadwick, but still had unfiltered water and still drank from the river, which was the repository for its sewage.

Hamburg relied on its reservoirs to remove any impurities in its water. The head engineer of the waterworks observed that as river water flowed into the reservoirs, the heavier impurities would sink and the flotsam including "dead fish and other corpses" would be blown to the side by the wind. By taking water from the reservoir "at the right level in the right way," impurities, he asserted, could be avoided. Since the floating "foam and its contents [could] be easily removed by hand," filters were, in his mind, superfluous.

Small animals, from snails to eels, did make it into the water supply, but a zoologist who studied the multitude of organisms that lived in the city's water pipes concluded that most of them should "possess enough strength" to resist being pulled up into the vertical pipes that supplied the city's buildings. In other words, despite the remarkable findings of Pasteur, Koch, and Snow, many of Europe's scientists failed to fully grasp the deadly power of the invisible.

Here Koch had found a natural experiment similar to the one Snow had studied forty years earlier in South London. Several key differences allowed him to erase any doubts about the validity of Snow's findings. First, he could compare a filtered to an unfiltered water supply when both drew from the same source. Second, he had the tools to look for the causal agent. Third, and perhaps most important, the world was finally ready to listen.

Koch showed that the river water crawled with *Vibrio cholerae* and that the filters removed it. The impact of the epidemic followed the water supply so closely that the homes around Hamburger Platz, a single neighborhood in Hamburg that used water from Altona, survived the epidemic largely unscathed. The findings reaffirmed the importance of the cholera bacteria and the role of water in its spread. They also gave public health officials a clear plan for driving a stake through cholera's heart.

The use of sand filters to purify public water supplies became routine on the heels of the disaster in Hamburg. Never had there been such a stark demonstration of the efficiency of a public water supply as a mechanism for delivering death to every door. The death of an old idea, however, is a protracted and ungainly thing.

In 1893 Koch's laboratory received a strange request. Professor Max von Pettenkofer, director of the Institute of Hygiene in Munich, wanted some cholera. The fact that he wanted cholera was not remarkable. Koch's lab had maintained a culture of *Vibrio cholerae* since the outbreak in Hamburg and may have been the only source of the bacteria in Germany. Anyone wanting to study cholera would need a sample. The thing that made his request stand out was that Pettenkofer had no plans to study the microbe in his own laboratory. He was convinced that the bacteria growing in Koch's culture media did not cause cholera and saw no need to study it further. He did, however, have an experiment in mind.

Despite the fact that Pettenkofer was a leading critic of Koch, Georg Gaffky, one of Koch's colleagues, prepared a sample to send. Pettenkofer was, after all, the elder statesman of public health in Germany. Gaffky packaged the sample with exceptional care and sent it to Munich. The sealed vial within contained enough *Vibrio cholerae* to kill hundreds of people.

At seventy-four Pettenkofer was one of Germany's most senior scientists with former students in key positions throughout the country. Among his protégés was Hamburg's former medical officer Theodor Kraus, the man who had treated Koch so poorly. (Kraus was fired in the wake of the epidemic.) Pettenkofer was convinced that cholera in drinking water could not, by itself, cause the disease.

Soon after receiving the sample, Pettenkofer stood before a small audience in a lecture hall at his institute. He explained to them his long-held belief that agents released from damp soils were the real cause of cholera. Koch's comma-shaped bacteria might predispose one to the disease he told them, but in the absence of appropriate atmospheric and hygienic conditions, it was harmless. Then he held up the vial containing billions of the bacteria and told the audience that he was now prepared to offer definitive proof of his theory. He then put the flask to his lips, tilted his head back, and consumed its contents.

What thoughts passed through Pettenkofer's aging mind at that moment? If Koch was right, as the majority of German scientists now believed he was, he could expect to be seized by cholera in the next forty-eight hours with a case so severe that it would almost certainly kill him. Was Pettenkofer steeled by absolute confidence in his ideas? Could he have been so certain that a theory widely seen as obsolete could save him? Had he simply disregarded the evidence Koch had put forward following the outbreak in Hamburg?

Pettenkofer's demonstration might have been, in his mind, the most fitting way to resolve that doubt. For this tired old man who had launched his career with a string of brilliant discoveries as an organic chemist, being wrong and dead might well have been better than being wrong and alive.

Or had Pettenkofer tried to hedge his bet? Perhaps he had snuck off to an empty laboratory and heated the broth to kill the bacteria in an act of private desperation. Perhaps a protégé seeking to protect his mentor had done it for him. It has even been suggested that Gaffky, learning of Pettenkofer's intentions, attenuated the bacteria before sending it out of respect for a foolish but honorable old man.

In the end it seems that someone had attenuated the culture, that the bacteria had not survived the trip from Berlin, or that Pettenkofer had been infected in the past and had some immunity. Current research suggests that the bacteria might simply have lost its virulence during years in culture. For whatever reason, after a mild case of diarrhea, Pettenkofer and his idea survived. His survival gave a last gasp of life to his theories, but in the years that followed, science trudged on toward the truth. The evidence that drinking water could spread cholera and other diseases grew steadily more clear. It is hard to pinpoint the precise moment the stubborn idea that microbes in water could not transmit disease finally died, but none marked the end of that centuries-old notion as starkly as the day that Max von Pettenkofer, nine years after his bizarre self-experiment, put a pistol to his head and blew the failed theory from his toothless skull.

8

THE SCRAMBLE
FOR PURE WATER

The spring mud sucked at the hooves of the horses trotting along Lake Street as a shrill whistle pierced the morning air. The sound came from deep within the foundations of the buildings that lined one side of the street. Carriages slowed to watch as a man with a trim goatee and thick black hair combed straight back paced in front of the imposing stone and brick facade of Chicago's Marine Bank, examining it with the voracious eyes of a twenty-eight-year-old entrepreneur. Eight large brick buildings, including some of the city's finest stores, stretched off to his right. Across the street a crowd had gathered to watch. Satisfied that everything was ready, George Pullman put his whistle to his lips and blew.

In the damp world hidden beneath the buildings, six hundred men scurried into action. Following Pullman's elaborate choreography, each man moved among his ten assigned jackscrews, pausing to give each one a quarter turn. Massive wooden support beams groaned and creaked as the jacks dug into their grain. Above them the ornate stone and brick buildings inched their way into the sky. When all six thousand jackscrews had been turned, each man returned to his assigned position and waited for the next pair of whistles. On the street level, bankers managed Chicago's money and the city's elite continued to shop unperturbed while above them printers, bookbinders, cobblers, and apothecaries went about their business knowing only that they would have a longer step to the street at the end of the day. In a masterful effort, the house mover from Albion, New York, would succeed in lifting a third of a city block five feet in the air, a tenth of an inch at a time. But Pullman had not come all the way from New York just to raise a few buildings. He had come to help raise an entire city.

The year was 1860. Chicago had just lost six percent of its population to typhoid, cholera, and dysentery. In response the city hired Ellis Sylvester Chesbrough, who had designed Boston's water distribution system. His assignment was not to design a water supply but, in the spirit of Chadwick and the sanitarians, to design a sewer. After arriving in Chicago and studying the lay of the land, Chesbrough immediately realized that he had been asked to plan a sewer for a swamp. If he dug under the streets and built sewers with enough slope to reach the river, they would enter the river well below its surface and, when the river was high, sewage would fill the basements of Chicago.

Chesbrough had a remarkable knack for bold innovation. After his father's business failure forced him out of school at the age of nine, he took whatever work he could find, but the job that defined his career was his work for a railroad contractor. There, in a time when such a thing was possible, he became an engineer with nothing more than an elementary-school education and job experience. His real-world education had bred in him an understanding of solutions that worked. With years of experience building railroads, he had no fear of massive engineering projects. So when he was faced with the prospect of building a sewer in a city that defied its proper construction, he came up with a radical plan. He would design and build sewers that worked and then reshape the city to fit the sewers.

In 1855 he began to lay pipes down the center of Chicago's streets, not by burying them, but by setting them on the surface. Using mud dredged from the Chicago River, he then built the streets up until they were higher than the pipes. This gave building owners a stark choice. They could either hire Pullman or a similarly specialized contractor to raise their buildings four to ten feet in the air or carve entrances into their second floors and watch as their first floors became basements. With the city redefined to meet Chesbrough's specifications, Chicago's sewage began to flow into the river.

For Pullman raising Chicago was just a prelude to a career that would earn him a vast fortune in the luxury railcar business and infamy as a ruthless manipulator of workers. Chesbrough and Chicago were also just getting started. The city's remarkable self-levitation helped clean the city, but Chesbrough's solution became his challenge. The intake for Chicago's water supply was six hundred feet out in Lake Michigan at a point about one mile from the mouth of the Chicago River. So the more efficiently Chesbrough drained the streets, the more efficiently the flow from the river contaminated the city's drinking water. A mile, it soon turned out, was not far enough. Again and again, cholera and typhoid rode out into the lake from the Chicago River, crawled back through the city's water pipes, found new victims, and slithered back through the sewers into the river to complete their circle of death.

Chesbrough recognized from the start that this could be a problem, but when he was hired, drinking water was not in his job description. That was the province of a separate agency. But after he had succeeded in draining the city, Chesbrough was given control of Chicago's water supply and charged with the task of providing clean water to the city. One can wonder whether this was a promotion or punishment, but in either case he once again hatched a bold plan.

In the winter of 1866, a team of Irish laborers huddled together as icy water splashed over the bow of their boat nearly two miles out in Lake Michigan. Below the deck a pair of mules struggled to maintain their footing in the rolling seas. Ahead, the men could see waves crashing against an immense stone pentagon, ninety feet across and topped by a forty-foot-tall lighthouse. Above the lighthouse an American flag flapped in the stiff breeze. They sailed up to one side of the man-made island, disembarked, and entered to begin their workday, completing what must stand as the most unique commute in Chicago history.

Inside the immense structure, a winch lowered two of the men and a mule down a narrow steel shaft that passed through thirty feet of lake water and then another forty feet into the lake bed to a brick-lined tunnel. The two men crawled out of the elevator and hunched slightly as they led the mule down the tunnel. They trudged over a quarter of a mile into the darkness, the light of their lanterns swaying with each step. The tunnel ended abruptly at a wall of dense, blue clay. The two men stopped and began to dig. Working in shifts for four dollars a day, the team would fill carts of clay, haul them out of the tunnel with the mule, and send them up on the elevator. After sixteen hours a team of bricklayers would take their place and spend the next eight hours lining the freshly dug tunnel.

At the same time, another elevator on the north side of Chicago carried a team of men to a similar tunnel seventy feet below the shoreline of Lake Michigan where they began to dig in the opposite direction. According to Chesbrough's plan, both teams would dig until they met in the middle. When the last brick was in place, valves would open within the stone pentagon filling the tunnel with the cold, fresh, and, hopefully, clean water. Steam-powered pumps would pull the water to shore. But first they had to make sure the two blind tunnels would meet.

Chesbrough's experience in the railroad tunnels of the East proved critical. With no way to tell direction underground, the laborers dug like moles. Engineers and surveyors made constant measurement, nudging the workers left, right, up, and down toward a spot on blueprint. An error of less than a tenth of a degree would send the two teams burrowing past each other. After almost two years of digging, they broke through the clay and the tunnels met. The planners missed by less than half a foot. The trivial error brought great relief to Chesbrough and his staff who had worried that the two tunnels might not meet at all.

Two weeks later a mule left the shore of Lake Michigan pulling a train of six cars laden not with clay, but with dignitaries (three men to a car) as two men pushed from behind. In eerie silence they creaked through the tunnel, staring at the parade of passing bricks in the dim flicker of their kerosene lanterns for almost half an hour. They stopped at a six-inch jog in the tunnel, which marked the spot where workers had met a few weeks earlier. In short order a similar train arrived after traveling east from the crib. There, in the surreal hollow of the tunnel, the men bent down to watch the mayor of Chicago lay the final brick in the tunnel and to listen as the echoes of his mercifully brief commemorative speech scurried off into the endless darkness.

A grander celebration of Chesbrough's success and a fuller measure of the public thirst for better water would come several months later as cheering crowds gathered to watch finely dressed regiments from the Chicago police, the Ellsworth Zoaves, the Masons, the Knights Templar, and the Dearborn Artillery (complete with horse-drawn brass cannons) march up Clark Street. With the memories of the Chicago Fire still fresh, the link between fire and water was powerful. In the heart of the parade, sixteen teams of powerful horses pulled fire trucks with names like "Liberty," "Economy," and "A. D. Titsworth," all belching steam into an icy March wind.

Halfway to its destination, chaos threatened to consume the parade. At the very spot where Pullman had raised the Marine Bank seven years earlier, a gridlock of wagons, carriages, snorting horses, and angry drivers blocked the parade. High spirits spiraled into nineteenth-century road rage, but a path was cleared and the marchers continued on to the newly constructed water tower.

Water had just begun to flow into the city and every petty bureaucrat who had ever signed a purchase order for a trowel was on hand to receive his due. As cannons fired, city officials droned on through a program that seemed to give every one of those officials, in order of ascending importance, his moment on the podium. Following the mayor's requisite speech, the crowd demanded to hear from Chesbrough. He rose only to acknowledge the graciousness of the mayor and the providence of the deity. The modest Mr. Chesbrough who had just engineered the longest underwater tunnel ever built had no speech prepared.

Chesbrough may have understood better than his audience that safe water is not an end, but a process, an ongoing struggle in which improvement is always possible and usually necessary. Purity, as it turns out, is fleeting.

At the same time that the unfortunate immigrant laborers were mining mud beneath Lake Michigan to build Chesbrough's tunnel, America's railroad barons were spinning a web of tracks across the newly settled west. The port of Chicago, at the center of the web, boomed. It would soon be the busiest port in the United States, serving more ships than New York, San Francisco, New Orleans, Boston, Baltimore, and Philadelphia combined. Manufacturing burgeoned. The population exploded. With each new production line and each immigrant to the city, demand for water grew and production of sewage grew in lockstep. To complicate matters further, a new sort of immigrant had begun arriving in Chicago in numbers never before imagined.

On December 26, 1865, a Burlington train pulled up to unload the first occupants for a vast new complex of buildings that sprawled across the prairie five miles south of Chicago. They clattered out across a wooden road, one of the finest in the city, exhaling clouds into the Midwestern winter. Another train followed, then another, then another. Before the next Christmas, a million and a half cattle, pigs, and sheep had arrived on similar trains to meet their fate in Chicago. By 1870 the annual influx had risen to 3 million and by 1900, 82 percent of the meat in America came from the jungle of meatpackers that grew up around the yards.

This vast city of animals brought with it many of the same problems as the human city to its north including the sewage that poured from the Union Stockyards into the South Fork of the Chicago River. That small tributary became the recipient of all things undesirable from the stockyard and its environs. At times the dried crust of sewage on the river's surface was thick enough to support a man's weight. Before long the bubbles of gas that rose from the fermenting accumulation of animal manure, carcasses, and human waste on the river's bottom gave it a new name. The South Fork of the Chicago River would henceforth be known as Bubbly Creek.

The stage had been set for the next waterborne disaster. It was only a matter of time and weather. As Chicago grew the vile waters of Bubbly Creek and the city's sewage reached out farther and farther into the lake until they began to choke Chicago's water supply. In 1873, with water levels low, cholera struck again, but other diseases had already begun to pose a greater threat. Cholera came occasionally and did not stay long. Improvements in Europe's sanitation and water supply were reducing its chances of crossing the Atlantic. But dysentery and typhoid fever could linger through the winter. These diseases became endemic. They came and stayed. And killed.

By 1891 Chicago had a rate of typhoid higher than any major city in Europe or North America. Afraid that the city's reputation as a center for endemic typhoid would scare away visitors, the organizers of the Columbian Exposition in 1893 laid a hundred-mile-long pipeline to bring in spring water from rural Wisconsin and installed a plant to distill any city water used on the fairgrounds.

Something had to be done. Chesbrough was gone, but the city fathers hatched a plan worthy of his legacy. Chicago saved its grandest, most audacious trick for last. When this project, at that time the largest civil engineering project ever undertaken, reached completion no dignitaries spoke, no bands played, and no cannons fired. Instead, at the last moment, a select few received hushed invitations to a christening held under the cloak of secrecy and darkness.

In the dim predawn of January 2, 1900, B. A. Ekhart stopped his carriage at the intersection of Kedzie and 35th Street and jumped out into the bitter cold with seven shovels, dropping them on the frozen mud face of a dam. One side of the dam formed part of the bank of the Chicago River. On the other side of the dam, the most remarkable of all the water projects in Chicago's history was almost ready.

The six other trustees of the Sanitary District soon joined Ekhart and the seven men rushed down the dam with their shovels. After scraping up a few token spades of frozen ground, they turned the task over to Dredge No. 7. The air throbbed with sounds of the powerful steam engine as the massive shovel began to claw through the dam. Eight feet from its goal, the dredge stopped, unable to reach the remaining wall of frozen clay and ice that blocked the path between the river and the Chicago Ship and Sanitary Canal.

The secretive opening ceremony stretched on for anxious hours. When four large charges of dynamite failed to breach the dam, the trustees grabbed their shovels and tried to finish the job by hand. When that failed they set fire to the wooden structures on the dam in hope of melting it open. In the end, workers managed to reposition Dredge No. 7 and the teeth of its huge bucket scratched through the final frozen jumble of ice and earth. Water began to seep into the canal. It would take more than a week for the canal to fill.

Once the vast canal that lay hidden on the far side of the dam was in operation, it would radically redefine the watershed for Chicago, its rivers, and Lake Michigan. The idea had first occurred to the French explorer Louis Joliet soon after he became the first white man to see the region in 1674. He recognized that a barely discernable ridge just ten miles west of Chicago separated two of the most important watersheds in North America. On the east side, all water flowed toward Lake

Michigan and ultimately the St. Lawrence River. On the other side, all water flowed west toward the Mississippi. Joliet imagined a canal cut through the ridge that would allow him to travel from Chicago to the Gulf of Mexico. The idea was simple, but 325 years would pass before the convergence of knowledge, ambition, technology, money, and necessity brought it to full fruition.

There had been several efforts to create a canal, but the resulting waterways were small, shallow, and ultimately inadequate. The project that was to be completed at the opening of the new century had no such failings. The Chicago Ship and Sanitary Canal was longer and wider than the Suez. It would redefine the course that the Chicago River had followed since the last ice age, moving it into an entirely new watershed. In an instant the river would begin to flow backward. The lake, which had been fed by water from the river, would now pour water into the river, taking with it the sewage of Chicago and carrying it into the Illinois River, across the state to St. Louis, and down the Mississippi to New Orleans.

The canal would be a great boon for Chicago, not only providing a permanent separation of its sewage and drinking water, but also making the city the connecting point between two great watersheds. Cities on the Illinois such as Joliet, Peoria, and St. Louis would have access to the canal, but they would also find themselves suddenly downstream from Chicago and on the receiving end of its raw sewage. Given that Chicago had recently recorded the highest rate of typhoid fever in the country, these cities were less than delighted at the trade-off. Fears that this displeasure might inspire lawsuits motivated the trustees to embark on their furtive midnight opening of the canal.

But even after they breached the dam, the canal's future was uncertain. Unless the city received clearance to open the gates at the far end of the canal in Lockport, they would have nothing more than the world's largest, most expensive, and most polluted swimming pool. As the canal slowly filled, the future recipients of Chicago's feces rushed to the courthouse. But which courthouse? Missouri courts had no jurisdiction over a canal entirely within the boundaries of Illinois. There was no hope that the courts of Illinois would help them stop the canal. The power brokers from the city of big shoulders could manipulate the Illinois legislature and legal system to their own ends. In a case without precedent, St. Louis could only turn to Washington for help. The trustees of the Chicago Sanitary Commission had no intention of letting that effort succeed.

So just after midnight on January 17, as the attorney general for the state of Missouri sped by train toward Washington to seek an injunction from the U.S. Supreme Court, the trustees of the Sanitary District of Chicago slinked onto a train to Lockport. Once again they hoped that the cover of darkness would shield them from the opposition. In the midst of a midwestern winter night, with the canal full, they opened the gates and Chicago's wastewater seeped into the Illinois River on its way to the Mississippi. They hoped that it would be far harder to stop the canal once it was in operation.

The legal battle would roll on for decades and would spark an international confrontation with Canada as water levels throughout the Great Lakes dropped almost half a foot, but the gates remained open. Chicago got its canal and with it a clean source of water for years to come. Typhoid, which had killed almost 2,000 people in Chicago in 1891 (far more than the Great Chicago Fire), killed only 322 in 1908 despite a doubling in population. Ultimately Chicago would treat its sewage and would build the largest water treatment plant in the world, but Chicago owes much of its success as a large city to a two-mile-long tunnel in blue mud and a river that runs backward.

Sitting on the largest contiguous body of liquid fresh water in the world, Chicago and other cities on the Great Lakes had alternatives unavailable to most other cities. Few cities could simply stretch intake tunnels miles out into a relatively pure natural body of water and perhaps no other city in the world could exercise such complete control of the sources of contamination that threatened its water.

Far more common was the situation faced by cities like Philadelphia, Washington, D.C., St. Louis, and New Orleans. Drinking from the America's great rivers, these cities inherited sewage from growing populations upstream. Major European cities on the banks of the Rhine, the Seine, the Danube, and the other grand rivers of the Old World had long faced this challenge and most had long since built filtration plants. On the heels of Robert Koch's discoveries and the epidemic in Hamburg, those cities that didn't filter were rushing to do so and many of those that already had filtration plants were improving them.

As American water supplies grew turbid and the consequences of that pollution grew apparent, many cities chose to follow the lead of the Europeans and install huge water filtration plants. Others, however, took a new course. Unlike Chicago, these cities did not have a vast lake of fresh water at their door. Undeterred, they chose to build their own.

The first Europeans in America had found a country laced with pure rivers and streams. Even as the twentieth century approached, that pristine wilderness still existed in the minds of urban Americans. It seemed that if one turned from the city and walked far enough into the forest, one could always find another mountain stream. By 1900 Boston and New York had already undertaken vast engineering projects to build rural reservoirs and pipe the pure water to their burgeoning populations. Other cities made their own plans to redesign watersheds to suit their needs. Most of these projects were steeped in controversy, power politics, and financial sleight of hand. The same powerful thirst that had Chicago digging under lakes and reconfiguring watersheds unleashed powerful, irresistible forces in the search for water. The huge projects that followed rolled over towns and left thousands of workers injured, maimed, or killed in the name of progress. But it was a reservoir in New Jersey that forever redefined the treatment of drinking water.

• • •

Bill Hoar pulled a wool hat over his thick red hair, lifted the fifty-pound diving helmet, and lowered it onto his head as if it were little more than a derby hat. His assistant, John Dobson, reached up like an attentive grandfather to tighten the twelve bolts that held the helmet onto Hoar's rubberized canvas suit. The cool spring sky spit rain as their raft bobbed on the water of the Jersey City Reservoir. Hoar was about to climb down into the water when Dobson grabbed him by the shoulders and stared through the glass faceplate.

"You'd better be careful, Bill," he warned, shouting so he could be heard through the brass helmet, "this ain't no jowk." (Dobson had somehow picked up a cockney accent on the streets around New York's Fulton Street Market, where he had been born and raised.) In 1904 any dive was dangerous. Dobson had assisted Hoar on hundreds of them, but something about this one worried him. He could see Hoar laughing at him through the glass of his helmet as the hulking Swede sank into the silence of the steel gray water.

The joke in Hoar's mind as he descended may have been the failed schemes of the dam's contractor, failures that now brought the two men from New York to the reservoir for the second time. Those problems began when, with the reservoir half full, a valve controlling one of the dam's sluiceways jammed. Like two giant drains on either end of the dam, the sluiceways allowed the dam operator to control the level of the reservoir as if it were an immense bathtub. Buried deep inside the dam, the gate valve should have allowed him to control flow through the four-foot pipe. Once it jammed, the operator had no way to stop the cataract that screamed through the dam and rushed out its far side in an angry torrent.

The reservoir was leaking, but the surging water in the pipe made it impossible to repair the gate. The contractor could have simply drained the reservoir, but that could take a week or more, and filling it would take two more weeks. The project was already almost two years overdue and every day of delay was costing him a hundred dollars, over two thousand dollars in today's money.

To avoid further expense, the contractor had designed an enor-

mous two-and-a-half-ton wooden ball to solve the problem. He had
to go 140 miles to Troy, New York, just to find someone who could
machine a ball five feet in diameter and fill it with lead. He planned
to lower the ball onto the pipe like an enormous bathtub plug that
would shut off the water. With the pipe empty, a mechanic could easily climb into the sluiceway and fix the gate. But in the damp chill of
early April, nothing was going according to plan.

A train had delivered the massive plug just before Easter. The following Monday morning, a steam-powered crane lifted it and swung
it out over the water. As the crane began to lower the ball into place,
the harness snapped. With a single explosive splash, the reservoir swallowed the plug. It plummeted to the bottom and dug into the mud of
what had been, just two weeks before, a farmer's field.

Hoar and Dobson had made their first trip to the reservoir to rescue the ball from uselessness. Hoar succeeded in reattaching the harness and the crane had raised it to the surface. His job complete, Hoar
returned to his small apartment on the Upper East Side of Manhattan.
He lived alone, but the work paid well enough that he could support
his sister who lived just across the East River in Queens.

The contractor made a second attempt to lower the ball into
place. This time the harness held as the crane operator eased the ball
onto the sluiceway. The water in the twelve-mile-long reservoir rushed
toward the four-foot-wide pipe and sucked the immense ball toward
its intended perch. The contractor stared intently at the cable that
held the ball. The cable slackened and, sure the plug was now in place,
he rushed to the other side of the dam and looked down its face. Below
him he could still see a huge jet of water escaping from the sluiceway.
He watched, waiting for the flow to stop. When it didn't, his heart
sank. The ball had failed to seal the pipe.

The contractor could only guess why. Perhaps the initial fall had
damaged the plug so that it was no longer round. Perhaps the ball had
never been perfectly spherical or the pipe itself was misshapen. A stone

on the bottom might have kept the great ball from settling properly into place. The answer was only a few hundred feet away, but it might as well have been on the moon. He could only stare at the rising water of the reservoir and guess. To solve the problem, whatever it was, he would need help.

So once again Hoar and Dobson rode the ferry to New Jersey with a pair of trunks laden with hundreds of pounds of equipment. Introduced to diving by his uncle, Hoar had become one of the strongest and finest helmet divers in the city. He enjoyed the close brotherhood that forms among men who share a common danger. He and his colleagues spent their work days walking along the bottom of the murky rivers around New York, searching for the lost, repairing the broken, and assisting in the construction of the bridges and piers that studded the city. With relatively crude equipment made only from wool, cotton, brass, glass, and natural rubber, he had come to feel at home in the crushing darkness of deep water.

After a quick lunch, Hoar and Dobson had rowed out to their raft and prepared to dive into the reservoir, charged once again with rescuing the contractor from his own incompetence. This time the challenge was far more dangerous than just pulling a giant ball out of the muck, but Hoar laughed off Dobson's warning and dropped into the cold. As he approached the bottom, the waters that were gathering behind the dam extinguished the last bits of sunlight.

Hoar felt his way along the bottom, toward the sluiceway. As he approached the ball once again, he could feel the rush of water escaping through gaps between the plug and the edge of the pipe. The strong current pulled at his legs. Careful not to get swept up in the flow, he searched for the reason the ball had failed to seal the pipe. The hazards that swirled around him in the soundless murky depths concentrated every fiber of his being on the ball and the water that rushed beneath it.

Up above John Dobson waited, wordless. His world had grown

as narrow and focused as his partner's, seventy feet below. He held Hoar's lifelines as he had on dive after dive over the past eight years, listening with his fingers for any signal from the bottom. The massive granite-faced dam, a hundred feet high and over three thousand feet wide loomed over him. His tiny raft bobbed like a cork as a pair of laborers cranked Hoar's compressor. They worked in thirty-minute shifts to send air down a narrow rubber tube to Hoar's helmet. Then Dobson felt it. Four strong tugs. Hoar was ready to return to the surface.

As he climbed up onto the dock, Hoar had a plan. He told Dobson he would need some sandbags and waited while workers on the dam prepared them. Then with the weight of the sandbags and a diving suit that weighed more than a hundred pounds, he climbed back off the raft and plummeted back to the bottom. He stuffed the first set of sandbags into the gap, but the water still rushed around them. When he was done, Dobson pulled him up once again.

Almost no one in Boonton had ever seen a diver at work and a crowd had gathered at the dam to watch. An amateur photographer among the onlookers snapped pictures as Hoar prepared to descend. The crowd craned to see as he gathered more sandbags and climbed back into the water. Once the water closed over him, all that remained were the bubbles that rose steadily to the surface and the constant efforts of those charged with keeping Hoar alive. The crowd watched Dobson waiting by the edge of the raft, his hands gripping the lifeline that ran down to Hoar. They watched the laborers who worked in shifts to turn the huge steel wheel, which in turn drove the pump that sent air through the thick rubber hose down to Hoar's helmet. They watched the torrent of water rushing from the bottom of the sluiceway for any sign that Hoar was succeeding. They could only imagine what was happening below the surface.

Out of sight Hoar braced himself against the ferocious flow of water that raced through the narrow opening as he dropped the sandbags into place. Each time he carefully set one of his twenty-pound lead-weighted shoes on the bag and pushed it into the crack. He could feel the remaining current quicken as if the water was growing more desperate to escape as the gap narrowed. He felt his plan was working.

After almost four hours of diving, as he strained to force one of the last bags into place, it broke. His foot slipped and in an instant the torrent of water pulled his entire leg into the hole. As it did the ball shifted slightly. He immediately tried to stand, but could not overcome the tremendous force as 15 billion gallons of water pushed toward the opening. He threw himself against the ball, but even his formidable strength was no match for its overwhelming mass. Up above John Dobson strained to detect the slightest pull on the line, unaware that his partner was trapped in the cold, dark, silent world below.

The people of Jersey City had known for almost twenty years they had a problem. The Passaic River had been pristine in 1854 when the city first began to draw upon it as a source of drinking water. In the years since, Passaic, Patterson, and other communities upstream from them had grown from small villages into busy towns and small cities. They had added sewers that sent an ever-growing stream of infectious waste-water into the river with relentless efficiency. The growth of industry had added sludge acids from the gasworks, washings from the silk and cotton mills, and the effluent from hundreds of small factories into the river. As a final insult, the Jersey City water works sent its own wastewater into the Passaic River at a point just above its intake pipe, the final ingredient for the vile stew that would fill the water pipes of the unfortunate city. Concerns about waterborne disease in Jersey City were so widespread that the Spanish government singled it out, placing a quarantine on all ships sailing from its harbor in 1893. By 1894 a *New York Times* reporter could state with little risk of hyperbole that the people of Jersey City "are thoroughly satisfied that they have the worst drinking water anywhere in the United States."

Where could they find pure water? Jersey City could have chosen to filter the Passaic, but the ability of the evolving technology to purify such foul water was uncertain and, far more important, filters were expensive to build and operate. So Jersey City scoured the lands around it in search of a water supply. There were several options to choose from and a massive contract to award once the choice was made. Political intrigue and cries of corruption plagued the process from its inception.

With the scent of money in the water, contractors circled like sharks. After six contentious rounds of bidding, in a strange, New Jersey-esque turn of events, the final round produced only one bid. Patrick H. Flynn, the lone bidder, proposed to dam the Rockaway River and build a twenty-one-mile-long conduit to bring water to the city. He had bid in previous rounds, but this time, as the sole bidder, his price had risen by $400,000. Despite cries of foul and legal challenges, the mayor awarded him the contract.

The awarding of the contract did not end the controversy. Even before the precise location of the dam had been determined, problems and conflicts haunted the construction of the Jersey City Reservoir. By the time the first spade of earth was turned in 1899, the plan had been challenged four times in court.

Still the project moved forward. Even without a dam site, workers began to quarry stone for its face. As they began to pull the huge blocks of granite from the bedrock of New Jersey, a stone broke loose. The foreman screamed out a warning and six men scrambled to safety. James Antonio, a forty-three-year-old immigrant from Naples, was hard at work when, hearing the commotion, he raised his head. At that instant the rolling stone, which would otherwise have passed over him, caught his head and slammed into a second stone just behind him. His head disappeared in an instant so horrible that the hardened quarrymen turned away, unable to watch. After four months in Boonton, Antonio died with a silver watch and $116 in his pocket, almost half of it in Italian lire.

The dam would continue to keep New Jersey's lawyers and doctors busy. Flynn, who immediately turned the project over to a sub-contractor, had defaulted on his contract. Mark Fagan, the new mayor bent on ridding the city of corruption, brought in a new contractor, but problems persisted. Lawyers were busy preparing a new round of lawsuits as the dam approached completion. By the time of Bill Hoar's dive, more than twenty workers had been seriously injured and eight men had died.

Time and experience had etched John Dobson's face and hunched his shoulders, further shrinking his short, stocky frame. Even he was not sure how many years he had lived. During those many hard years, he had felt the cold breath of danger many times. As he leaned over Bill Hoar's lines, he thought he felt it again. His partner had been down for two hours after suggesting he would be done quickly. Then four tugs on the line brought him a moment of relief and he began to pull. He had pulled up just two feet of line when the rope stopped short. Dobson felt a sudden signal to slacken the line and then a signal to pull again. Dobson's fears returned in a dreadful rush. Something was wrong, but he had no idea what.

He called for help and workers rowed out to the raft to pull with him. As the team of sinewy construction workers pulled, the rope again went tight and Hoar once again signaled for them to stop. Soon it became clear that brute force was not enough. Dobson had to find help for his partner, just seventy feet away but lost in another world. His only hope was to get another diver. The closest one was hours away in New York City.

When the telegram for help reached the office of John S. Bundick, the diving contractor in Manhattan, the call went to William Oleson. As soon as he received word that his friend of seven years was in trouble, Olesen assembled his equipment and rushed to the ferry terminal. He made it across the Hudson just in time to miss the train to Boonton. Desperate, he explained his plight to the stationmaster who arranged for a locomotive and a single car to rush him to Boonton. Riding on a spur line that had been built to bring equipment and supplies to the construction site, Olesen arrived just two hours after the message was sent.

It was seven o'clock. Evening was falling. Olesen found Dobson on a raft crowded with the strongest workers from the dam site, along with the pump and its ceaseless operators. Their collective weight had sunk them waist deep in cold water as they struggled to free Hoar from his underwater prison.

William Hoar was not sure how long he had been down. Under normal circumstances a diver could expect to work for four hours at that depth, but time kept above the surface had ceased to exist. A cold, black vice had squeezed his universe down to the ball, the pipe, and his body. The remorseless ticking of his physiological clocks marked the only time that mattered. The infinite cold crept through his suit, past the layer of wool and into his flesh, driving his body temperature down. The pain in his leg surged through his body. Even the twin vultures of thirst and hunger had started to circle. He had been working hard with nothing to eat since morning. Even if cold, pain, thirst, hunger, or their combined effects did not take him, even if he could make his way to the surface, the nitrogen that lay hidden in his blood might well turn to deadly gas bubbles as he rose, killing him with the bends as the pressure dropped.

Hours passed in the surreal horror. Other than the steady waves of his breathing and the occasional distorted echoes of sounds from the surface, the silence was absolute. He pulled regularly on the lifeline to let Dobson know he was all right. In every sense the lines provided his only link to life.

Then he thought he heard something. Sounds scuttled through the darkness. Was his imagination playing tricks on him? Suddenly he felt a hand and then a mask pressed up tight against his. A voice echoed through his helmet. The words were muffled. "Bill, are you okay? What happened?"

The appearance of Oleson, one of the best divers in the business, must have brought new hope to Hoar. Hoar's confidence rose. Shouting so he could be heard, he explained that his foot was caught in the current. Oleson could feel its tremendous force as he bent down, hoping to remove the sandbags around Hoar's foot.

Oleson worked quickly. By the time he had been able to assemble a new raft and get down to Hoar, it was well past nine P.M., more than eight hours after Hoar's initial descent. He broke open the sandbags and the sand disappeared down the sluiceway, but Hoar remained pinned by the ball and the current. Then a sudden, horrifying rush of icy, black current grabbed Oleson's hand and sucked it into the gaping maw of the pipe.

In an instant Oleson found himself next to Hoar, his hand caught between the ball and the steel rim of the sluiceway. The rescuer had become fellow victim. He pulled at his hand with all his strength. After a long moment of terror, he felt a searing pain as his hand popped out of the trap. He fell backward holding his bleeding hand above him.

Oleson returned to the surface to consider his options. They were few and poor. On the other side of the dam, workers had tried to plug the sluiceway. If they succeeded, the water would rise, the current would stop, and Hoar could escape. Despite their efforts, however, water continued to roar through the pipe. Oleson could also try to move the ball. It might be possible with a proper harness and a team of horses on the shore, but the sudden rush of water would be almost certain to take hold of Hoar, pull him down the pipe, and crush him against the broken gate. Oleson even considered an underwater amputation, but did not believe his friend would reach the surface alive with a severed limb. It seemed that his only option was to drag the poor man to freedom.

Despite his own wounds, Oleson worked through the night to

free his friend. He brought down a block and tackle so that work-ers on the surface could apply more force in their effort to pull him loose. When that failed, he rigged a complex compound pulley system with one end connected to weights on the reservoir bottom. Workers on the surface pulled on the rope, but instead of moving Hoar, their combined force simply moved the weights. Oleson added more and more weight, but only succeeded in dragging a 350-pound anvil and a pile of sandbags through the mud. Hoar remained trapped. Somehow Hoar maintained his strength and his wits. Through it all he leaned up against the ball, moving to one side and then the other to assist in his rescue.

Word of the crisis had reached Boonton and the crowd swelled. Many stayed through the night as Oleson dove down again and again into the pitch black. Ten times he descended. Ten times hope rose. But as the moonless night wore on, hope and the crowd slipped off into the gloom.

By the time dawn crept across the reservoir, Oleson was nearing his breaking point. He descended one more time, taking with him the end of a massive rope, which he fastened to Hoar's waist. Before leav-ing, he took his friend's hand, squeezing it to reassure him. Hoar had weakened but managed to squeeze back. Oleson returned to the sur-face and collapsed. As a doctor attended to him, workers on the shore, nine hundred feet away, attached the other end of the rescue rope to a team of horses.

The teamster called out, the horses pulled, and the rope drew taut. It inched shoreward. A final glimmer of optimism ascended and then collapsed as the rope went slack and the horses stumbled forward. Somewhere below the surface, the rope had snapped.

The pull of the horses jerked Hoar up and his ankle exploded in pain. His suit tore away, and cold water rushed up his leg. It must have seemed as if the force of the horses would rip him to pieces. Then it abruptly stopped, leaving him clinging to life by a thread. The hole in his suit had created a new peril. The air from above, from the men who had pumped life to him all night long, would escape through the hole in his suit if he fell over. If it did black, suffocating water would rush in, bringing certain death. Down to his last ounce of strength, he struggled to stay upright.

At one P.M., twenty-four hours after Bill Hoar's first descent, John Dobson let out a plaintive cry. He had been there each minute of those twenty-four hours, making sure the pump continued to run, waiting for a signal.

"What is it?" called the workers on the dam.

"I felt a tug on the line," he replied.

Dobson shook the line in response and was sure he felt Hoar shake the line three times. Perhaps the old man was just imagining things. Or perhaps, with his last bit of strength, Hoar was saying good-bye to an old friend.

Two more days would pass before a diver could bring Hoar's body to the surface. That was possible only after another diver who was also a mechanic had descended through another shaft in the dam to repair the broken gate. The sluiceway filled, horses pulled the ball to one side, and workers pulled Hoar's remains up onto the raft. As the helmet was pulled off, surface air seeped in for the first time in almost one hundred hours, and the team of men who had pumped air to him constantly for four long days and nights ceased work. A doctor stood nearby, but the sight of his blood-suffused head confirmed everyone's worst fears. The men who had gathered around him on the raft could see where the struggle had stripped the skin from his ankle and foot. The following Sunday, four divers carried Hoar's casket down the steps of Our Lady of Mt. Carmel Church in Astoria, Queens.

Hoar was just one of many casualties in the scramble for pure water. Massive engineering projects to redirect rivers, reshape watersheds, and create vast new lakes relied on new, untested methods and required vast numbers of workers. Workplace safety rules were almost unheard of. Hundreds of men died to slake the rapacious thirst of America's growing cities. But it was not just individual workers who died. Entire towns perished in the quest for pure water.

The Jersey City Reservoir drowned the village of Old Boonton and part of Parsippany. Katonah* disappeared beneath the waters of New York City's Croton Reservoir and when that growing metropolis exhausted the waters of Westchester, it stretched north to the Catskills, consuming the towns of West Hurley, Ashton, Glenford, Brown's Station, Olivebridge, Brodhead, Shokan, West Shokan, and Boiceville beneath a huge reservoir almost ten times the size of Jersey City's. Boston's vast Quabbin Reservoir devoured four New England towns when it flooded the Swift River Valley. Even today one can still see the ghost of Dana, Massachusetts, staring up from beneath the waves.

Each town fought back, but few had a chance against the large cities that laid claim to their land. The struggle of Old Boonton was only a minor skirmish in the story of the Jersey City Reservoir. Far greater controversies surrounded the great stone dam through its construction, but the most important one came long after the dam was complete. That story would change the history of drinking water.

In the spring of 1904, the dam operators opened the valves that sent the water from the Rockaway River through twenty-three miles of cast-iron conduit to the taps of Jersey City. After the initial relief that the controversial project was complete and the vile waters of the Passaic no longer violated their faucets, a controversy began to brew. Proponents of the project had carefully ignored the fact that twenty thousand people lived in the Rockaway River watershed above the reservoir, many of them in the town of Dover. The rush of sewage that entered the reservoir during heavy rains made this studied ignorance difficult to maintain.

---

* Some of the towns mentioned were relocated and exist today at new locations.

Mark Fagan, the mayor of Jersey City, had come to power in the midst of the dam controversy by an unprecedented defeat of the local political machine. He had already taken on Patrick H. Flynn, the dam's unctuous first contractor. He was not about to pay for a reservoir full of dirty water. The son of poor Irish immigrants, Flynn had been raised to view life as "one long fight for what's right."

In October 1906, with the controversy at a bitter stalemate, an outbreak of typhoid claimed fifty-three lives in Jersey City in a single month, three times as many as in all of 1905. When the superintendent of the Bureau of Contagious Disease ventured up the Rockaway in search of the outbreak's source, he found three cases in the town of Dover. When he went to the town hall, however, local officials refused to give him information about the cases. But those officials could not conceal the fact that Dover discharged its sewage into the Rockaway River.

News of typhoid in Dover turned the controversy in Jersey City into a pitched battle. The contractor had powerful friends and they turned on Fagan, insinuating that the outbreak was a fabrication intended to further his position in the fight over the dam. Physicians appeared out of the woodwork to state that the water was untainted. Others suggested that the diagnosed cases of typhoid had been incorrectly identified. "Concerned citizens" appeared outside the city hall protesting the mayor's irresponsible actions and their negative impact on the city's image. When the Board of Health prepared flyers for schoolchildren warning them of the typhoid risk, the city's police chief, a stooge of the machine, served notice that he would arrest any board member who attempted to distribute them.

Seasoned in the rough streets of Jersey City, Fagan feared no one, not even the machine. He refused to back down from a mere contractor. He insisted that the contractor solve the problem by treating the sewage from Dover and Rockaway. The contractor, in an effort to avoid that expense, looked for another solution. He found it in Chicago on the banks of Bubbly Creek.

• • •

For most of Chicago, the remarkable reversal of the Chicago River carried off the problems that had faced its water supply. Rates of waterborne diseases dropped dramatically. But for one large segment of its population, the problems got worse.

The reversal had failed to flush out Bubbly Creek. Instead it turned the small fork of the Chicago River into a nightmarish backwater. For people living close to the creek, this meant enduring a terrific stench. For livestock at Union Stockyards, which relied on the creek both as a repository for waste and a source of water, this meant endless waves of disease. The fact that this translated into huge financial losses and threatened the viability of the yards brought George Johnson to Bubbly Creek.

Johnson was a consulting engineer and the owners of the stockyard charged him with the task of providing safe water for the livestock. Cost was, without question, an issue. Using the Chicago River would require a long pipeline, and Chicago's sewers, which had once been downstream, were now upstream from the yards. Lake Michigan was simply too far away. That left Bubbly Creek as the only available water supply, but it was profoundly polluted. To purify enough water to supply the vast transient herds was far too costly. No conventional option seemed feasible.

Johnson took a radical new approach. Rather than filtering the water, he would disinfect it with chlorine. Chlorine had been tested in a few small plants and had been applied to sewage as a disinfectant, but it had never been used to treat water on this scale. As Johnson supervised the construction of the plant, he got a call from Jersey City. The beleaguered contractor for the Jersey City Reservoir had heard about his project. If Johnson's system was good enough for cows and pigs in Chicago, he wondered, why wouldn't if work for the people of New Jersey?

By the summer of 1908, Johnson had built a treatment plant near the outlet of the Jersey City Reservoir unlike any other plant in the world. Inside three tanks, each the size of a small swimming pool, contained 10,500 gallons of a substance then referred to as chloride of lime, the same white substance that the General Board of Health had encouraged communities to spread along the streets of London fifty years earlier to slow the spread of cholera. The contractor turned on the plant and returned to Mayor Fagan, demanding that he be paid.

Jersey City sued, insisting that the contractor was obligated to provide water free from the upstream sewage. The contractor insisted he only needed to supply safe water and had done so. In the end the courts agreed. In December 1908 the *New York Times* acknowledged what would prove to be one of the most important public health interventions of the twentieth century with four short paragraphs on its back page. Sandwiched in between articles about a threatened boycott of automobiles built in New Jersey and the appointment of a fourth member to the New Jersey Fish and Game Commission, was the paper's report on the first routine use of chlorine for a municipal water supply.

"So successful has been this experiment," the article stated, quoting an unspecified source, "that any municipal water plant, no matter how large, can be made as pure as mountain spring water." The article went on to note that New York had decided to add chlorine to the waters of the immense Ashokan Reservoir which was still under construction in the Catskills. Indeed, water suppliers all over the country had been looking for just such a technology. Those that did not filter leaped on a treatment system that required a small fraction of the capital and space needed to construct a filtration plant. At the same time, operators of filtration plants needed a method to protect against the pathogens that inevitably find their way through a filter bed.

Engineers around the country and in Europe had experimented with other ways to disinfect water including ozone and ultraviolet light, but nothing provided the low cost and ease of use associated with chlorine. Within six years half the water treatment plants in the United

States were using chlorine to disinfect some 2 billion gallons of water each day. By 1924 three thousand cities had turned to chlorine (mostly in the form of chlorine gas) to treat almost 4 billion gallons of water.

Employing chlorine together with rural reservoirs, filtration plants, or both, municipal water suppliers around the country soon had systems that routinely provided water free from disease-causing bacteria. Serious waterborne diseases like cholera, typhoid, and amebic dysentery dropped from routine to rare. In 1900 an average American had a 5 percent chance of dying of a gastrointestinal infection before the age of seventy. By 1940 that rate had dropped 0.03 percent and by 1990 it had fallen to about 0.00005 percent.

By 1960 drinking water had disappeared from the national consciousness as an issue of concern. Civil engineering had been the high technology of the nineteenth and early twentieth centuries, but the luster faded. With the opportunity to reengineer watersheds all but gone and the technology of treating water mature, providing drinking water became mundane. The primary goal for those maintaining public water supplies became not innovation, but invisibility. If you did your job right, no one would notice. If you were noticed, it meant the water smelled bad or tasted funny, or, heaven forbid, had caused a detectable disease outbreak. Success bred complacency. Soon complacency drifted into somnolence. Then, as the century wore on, problems began to disturb this slumber.

9

THE TWO-EDGED SWORD

The cold nights of autumn had already begun to paint the leaves of Boston as I walked from Vanderbilt Hall past the classic quadrangle of Harvard Medical School to the uninspired concrete monolith that houses the university's School of Public Health. Once inside I asked the security guard for directions to the offices of the Technology Assessment Group. To my left was a large bank of elevators. The guard, a man whose accent I could place only as from somewhere on the Indian subcontinent, gestured in the other direction past the kitchen toward a single small elevator next to a cleaning closet. That elevator went in only one direction, down. As the elevator descended past the basement and into the bowels of the Kresge Building, I felt sure the guard had misheard me.

I had arranged to spend two months during my senior year of medical school working with Tom Chalmers, a leader in clinical research and the former dean of Mount Sinai Medical School. Dr. Chalmers had pioneered the use of a new and important tool in medical research known as meta-analysis. After leaving Mount Sinai, he had paired up with Fred Mosteller, a chaired professor of mathematical statistics at Harvard, to create the Technology Assessment Group (TAG), which was devoted to the methods and application of meta-analysis. I had expected to find these two giants of medicine and statistics in a suite of offices overlooking the Boston skyline, but as I made my way past

the copy center to an oddly angled cluster of offices in the second basement of the Kresge Building, I understood that Harvard was not like
other places.

When I arrived, TAG's offices were empty except for a short man
with a graying shock of black hair, a day or two of stubble on his face,
and black-rimmed Coke bottle glasses. "Hi, I'm Bruce," he said with
a New York accent and a wry smile, "Bruce Kupelnick. You must be
Tom's new person." I would soon learn that Tom had many "persons."

Bruce showed me to my "desk," actually half of a small desk in
the middle of Tom's office. The other half belonged to another one of
Tom's people. Bruce took great pleasure in showing me around the
office, introducing people as they arrived. He had left the University of
Chicago in the midst of his doctoral dissertation in history with plans
to write a book. The book had proved to be far more elusive than he
had anticipated. He had found an ideal spot as Tom's assistant, embed-
ding himself in the unique intellectual buzz of Boston.

As people arrived the conversation turned to the success of the
Red Sox, who were in the process of winning the American League
East. Amid the gathering wonder as to how the Sox would manage to
blow it this year, a tall angular figure swept into the office. His thin-
ning gray hair had just a hint of its original red and appeared windswept, as if the sheer pace of his life were blowing it back. A matching
mustache bounced on his upper lip. Tom Chalmers smiled, stuck out
a long arm, and shook my hand as he greeted me with unaffected
enthusiasm. He invited me into his office, but not before adding his
thoughts on Boston's shot at the pennant.

Tom's office had a narrow desk facing the wall at a height that
allowed him to stand and work. He rarely sat. From it he held court
on the rare days when he was not on the road. (I counted eleven days in
the office during the two months I was there.) Tom's presence brought
a steady stream of visitors. One by one Tom's people came by to discuss
their projects.

For the moment I had an audience with the master. From his desk Tom produced a thick folder crammed with photocopied papers. I leafed through the dog-eared articles, as he explained that each of the papers examined the relationship between chlorinated drinking water and cancer. A student of his had assembled them for a course he taught on meta-analysis. She had not been able to come up with a system for combining the results of the papers. Tom suggested that my first project should be to see if I could figure out a way to make the study work.

To a student the "suggestion" of a research mentor has the weight of an edict from on high. I sat down at the fraction of a desk that had been carved out for me and set to work. I dug into the stack of papers in search of the pattern that would allow me to piece together the puzzle. I soon realized that I would need to burrow back to the birth of the idea that spawned them if I wanted to understand the papers. One of the best places in the world to do that was next door at Harvard's Countway Medical Library.

Every minute of every day, the world's researchers are adding to the ever-expanding inverted pyramids of medical knowledge. In the stacks of Countway, I excavated down to the well-worn bricks buried deep within the pyramid. As I did so, a larger story took shape.

On April 22, 1915, with dawn breaking outside the Belgian town of Ypres, German seventeen-inch howitzers suddenly opened fire. For thirty minutes French and Algerian forces huddled in their trenches, and then, abruptly, the guns fell silent. Carefully the men rose up and peered out toward the rising sun. They scanned the horizon expecting to see German soldiers running through no-man's-land with fixed bayonets. At first they saw only the sun rising over a barren landscape with a light breeze blowing from the enemy lines. Then something began to grow on the horizon. As they watched, a vast greenish yellow cloud began to slither toward them.

They could not know the nature of the demon that approached. The initial bombardment had provided the Germans cover while they unleashed 168 tons of deadly gas. The east wind carried the poison over four miles of trench lines where it tore at the lungs of some ten thousand soldiers, killing half of them. Some had survived by running from the cloud, but when the Germans advanced behind the gas, they encountered thousands of Allied soldiers coughing violently, temporarily blinded, and stumbling among the corpses of their countrymen.

The chemical that the Germans had chosen to launch the era of gas warfare was chlorine. The extreme reactivity of chlorine that made it a potent disinfectant in drinking water made it a deadly weapon in chemical warfare. At the time of its introduction for water treatment, the concentrations used in drinking water did not appear to have any detectable toxic effects in humans. Jersey City could not even find a scientist to argue that chlorination of drinking water might have deleterious effects on human health. The tremendous benefits associated with inactivation of harmful pathogens made arguments against its use seem heretical at best.

The level of chlorine used to disinfect drinking water is far below that needed to cause acute toxic effects like those inflicted by the German attack. In the decades following its introduction in Boonton, any ill effects of chlorine seemed hardly worth consideration when compared to the obvious benefits of disinfection. But its ability to react with and kill microbes would create an unanticipated problem. I would find the story in the papers that Tom Chalmers handed me on that October morning in Cambridge.

The first hint of a problem came in 1970, when Johannes Rook, a chemist for the Rotterdam Waterworks, filled a bottle with the city's treated drinking water. With massive waterborne outbreaks of cholera and typhoid a distant memory, water suppliers had moved on to other concerns. Among them was improving the taste of the drinking water, a major source of customer complaints, particularly when chlorine doses were high. Charged with this task, Rook had a problem; standard methods for finding chemical contaminants in drinking water were useless to him.

From his experience as a chemist for a Dutch brewery, Rook knew that much of the flavor in a glass of beer floats in the air above it, filling our noses before we even have the first sip. Once these volatile organic compounds (VOCs) have evaporated, the flat beer that remains offers little taste. In 1970 the standard methods for testing water failed to capture the VOCs. Rook needed to find a new way to test water. Undaunted he adapted a method he had used for testing beer. What he found when he used that method to test water shook the complacent world of drinking water treatment.

Using equipment of his own design, Rook collected the chemicals that evaporated from the water. He then separated these VOCs by injecting them into a gas chromatograph, a device used to help identify unknown chemicals. On one side of the chromatograph, ink flowed from a thin finger of metal onto a long strip of paper as it rolled past. As each different VOC passed through the chromatograph, the pen jerked, leaving a series of spikes, like a range of narrow mountains on the paper's red grid. The size of the spike corresponded to the amount of the chemical that was present. When he examined the graph, he was shocked. The largest spike corresponded to chloroform, a chemical he had never expected to find.

Chloroform was no longer seen as the benign chemical that John Snow had offered with such confidence to the queen of England. In 1945 scientists had shown that high concentrations caused liver cancer in mice and could also destroy their kidneys. Inadequate regulations had allowed its continued use in a number of products, but its presence in drinking water raised the stakes dramatically.

Where could it have come from? No industry in Holland could account for the amount of chloroform present. To find out, Rook began to test samples of the water coming into the treatment plant. The results were even more alarming than the initial discovery of the spike. The river water flowing into the plant contained no chloroform at all. Rook could draw only one conclusion. The process of chlorination was causing the formation of toxic chemicals including chloroform in the drinking water.

In the sixty years since its introduction, chlorination had become central to drinking water treatment. The possibility that, as it made the water safe from pathogens, it was introducing a new set of poisons had staggering implications. Rook checked with the plant's health officer who assured him that many cough medicines contained chloroform and its presence in drinking water should not be a source of concern. Nonetheless, Rook recognized that the news could be explosive. For the next four years, he kept his findings quiet as he conducted more tests.

In 1972, as Rook refined his study, a report from the International Agency for Research on Cancer upped the ante. The authors of that study pointed to the 1945 study, which, while small, clearly suggested that chloroform could cause cancer. More research was urgently needed.

Four years would pass before Gordon Robeck picked up his phone on October 15, 1974. As he did he wondered why a reporter from the *Miami Herald* would be calling him. Robeck was head of the drinking water treatment research group at the Environmental Protection Agency (EPA) and prepared himself for routine questions about pesticides or industrial chemicals in the water supply. But this reporter had a question he never expected. Somehow this reporter had heard that chlorination caused the formation of chloroform in drinking water. Was it true?

As it turned out, Tom Bellar, a chemist at EPA's drinking water labs in Cincinnati, had been investigating chemical contaminants in the city's water at the same time Rook had been analyzing the flavor of Rotterdam's water. Bellar had studied air pollution before shifting to water and, like Rook, brought a unique perspective to the chemical analysis of water. Faced with the challenge of developing a better method to measure volatile chemicals, Bellar had come up with a method very similar to Rook's. When he applied the method, he discovered, just as Rook had, that the treatment of drinking water produced an array of chlorine-containing compounds, including chloroform. For the four years that followed, the EPA and Rook had sat on the same explosive news.

To avoid release of the information, the EPA simply chose not to publish its findings. Rook followed a strategy that was almost equally effective. He published his findings in Dutch. Even though he included the graph from his gas chromatograph he did not mention that the large peak corresponded to chloroform. Only an astute reader of the Dutch literature would have understood the implications of Rook's work.

One such reader found his way onto the staff of Miami's pollution control program. In 1974 he mentioned Rook's work to a friend, Mike Toner, a young science reporter at the *Miami Herald*. Rook's and Bellar's discovery was about to detonate.

As Toner dug into the story, he placed the call to Robeck. Robeck and the EPA had been able to stay silent as long as nobody knew the right questions to ask. They had hoped to contain the story until they could find a solution to the problem. When Toner asked him point-blank if the chlorination of drinking water resulted in the formation of chloroform he had no choice but to answer "Yes."

The October 17, 1974, edition of the *Miami Herald* carried Toner's story. The national media picked up the story and by the end of the month NBC had begun a three-part series on chlorinated chemicals in drinking water. This was not the managed release of information EPA had in mind. But the story was just beginning.

• • •

Those called to epidemiology find themselves in the peculiar business of picking through recent history in search of public health mistakes. From cigarettes to fried foods, epidemiologists have shown us the fallacy in our assumptions about the safety of the things we eat, drink, and inhale. In 1975, when the EPA released a report showing that water supplies all over the United States were contaminated with a variety of chemicals including chloroform, environmental epidemiologists pricked up their ears.

The story turned ominous when, in 1976, the National Cancer Institute reported that long-term exposure to chloroform caused cancer at concentrations far lower than those studied in 1945.

Well-intended efforts to protect us from pathogens in drinking water had resulted in a natural experiment on a massive scale. Tens of millions of people had been subjected to a lifetime of exposure to a known carcinogen. Epidemiologists around the United States set about assembling and analyzing the results of that experiment. The thousands of hours put in by more than seventy epidemiologists ultimately produced the stack of papers that Tom Chalmers handed me in the fall of 1990. In just thirteen years, they had generated more than twenty-two studies, all designed to answer a seemingly simple question: Does exposure to chlorination by-products in drinking water cause cancer?

I knew that the individual papers must tell a mixed story. The purpose of the meta-analysis was to combine the results of different studies to make sense of that confusion. I spent the next week immersed in books and papers on the methods of meta-analysis, picking the brains of Chalmers and Mosteller and planning how to approach the challenge of combining the results of the studies. The Boston sky was often dark by the time I emerged from the windowless confines of our basement warren at the end of each workday. On those nights when the cafeteria was still open, I would sit and eat while reading a few more papers before climbing up to my small room on the third floor.

With nothing more than a bed, a chair, a desk, a lamp, and an alarm clock, the room suited the monklike existence I would lead during the two months I had to complete the project. Each night I plowed through a few more papers before collapsing into bed. Each morning I woke early for a run along the Fenway and a quick breakfast before crossing Longwood Avenue, a narrow canyon between the ever-growing cliffs of Harvard's medical empire. The early morning was my only chance to steal a quiet moment for work, so I made a point of arriving at the office before the crowd came to bemoan the latest pummeling of the Red Sox in the American League play-offs.

As the play-offs wore on, I worked with Bruce to make sure we had not missed any papers. Bruce was among the most literate people I had ever met. He once told me that he read some sixty books each month and was rarely without a book in his hand. Bruce was happiest in the musty stacks of the world's great libraries and archives in search of hidden treasure. Together we exhausted Harvard's Countway Library, tracking down every possible article.

The studies all asked the same question, but took widely different approaches to answering it. The challenge of meta-analysis is finding the common threads in the disparate analyses and using those threads to stitch them together in a meaningful way. Tom's office was a beehive of intellectual activity, but not always a great place to solve a puzzle. At one point, as I struggled to make the study work, I went to Fred for help. I told him I needed a place to think.

Fred's office was an orderly refuge amid the creative chaos of the Technology Assessment Group. He and Tom were the intellectual odd couple, two men with opposite personalities who had found common ground in the late autumn of remarkable careers. A man of few words, Fred hesitated a moment before opening his desk and handing me a key. "Why don't you use the hideaway," he said. This was a key to the world of the chaired professor, a world where one finds space in a place where there is seemingly none to be had.

I followed Fred's directions through the School of Public Health to a nondescript door that I had passed a dozen times without noticing it was there. Behind it was a large room, twice the size of most offices in the school. The walls were lined with bookshelves filled with carefully organized bound copies of the world's major statistical and biostatical journals going back several decades. In its center were the hideaway's only pieces of furniture: a small table and a wooden chair. On the table sat a neat stack of yellow legal pads and a box of pencils. Nothing else. I felt as if I had entered some sacred temple of the intellect.

There, in Fred's hideaway, I put together the pieces of the puzzle. As I assembled the data that Tom and one of his graduate students had pulled from the papers, a consistent picture began to emerge. A lifetime of drinking chlorinated water increased the risk of bladder cancer. It appeared that it might increase the risk of other types of cancer as well, especially colorectal cancer.

I got a hint as to the controversy I was about to stir up when I began to send the paper out for publication. I routinely received two responses: one would praise the paper and herald its scientific import and the other would attack it as an inappropriate use of meta-analysis. The editors, looking for two strong reviews, would reject it. In the summer of 1991, I found a way to get it published with the help of a hurricane named Bob and a remarkable woman named Devra Davis.

That spring I had been invited to serve as an adviser to a National Academy of Sciences Committee on Environmental Epidemiology. Devra Davis, a longtime advocate on issues related to environmental health, was then working for the National Academy and was in charge of shepherding the committee through the process of producing a report. The committee had already met several times in Washington when Devra scheduled a meeting at one of their other conference facilities—an old Cape Cod estate near Woods Hole. As I arrived at Logan airport, I noticed the headlines in the *Boston Herald*. "Bracing for Bob," it read in the paper's usual hyperbolic type.

Twenty-four hours later I huddled in our hotel with several of the committee members. The air cracked with the sound of huge old trees splitting into pieces. Across the street the storm surge rolled in, ripping boats from their moorings. The devastation went on for hours, interrupted only by a few ethereal minutes of calm as the hurricane's eye passed over our heads. With power lines down all over the Cape, the arrival of night plunged us into complete darkness. We awoke the next morning to find the shoreline littered with boats and the roads clogged with tree limbs. The power would not return for three days. Hurricane Bob was one of the worst storms ever to hit the Northeast.

The committee had no choice but to make do. We met in unlit rooms, ate food warmed over Sterno, and went to bed by candlelight. One day I presented my study to the committee and explained my difficulty in getting it published. Devra suggested I send it back to the *American Journal of Public Health*. The journal, she said, had a new editor who might reconsider a decision by his predecessor to reject it. I took her advice and the following summer it published my paper.

The reaction shocked me. Within twenty-four hours the National Cancer Institute (NCI) and the Environmental Protection Agency issued a press release denouncing the study. The NCI response surprised me since their own scientists had studied the risks of chlorination by-products with results similar to mine. I suspected that I had treaded on the turf of the EPA, but again I could not understand the vehemence of their response. It would take years before I could understand the strange world of science, politics, and money that I had just entered.

In August 1992 I received a call from a representative of the International Life Sciences Institute (ILSI). They were hosting a meeting on the safety of drinking water disinfection in Washington, D.C., and hoped that I would attend. I had been aware of the meeting, but had only recently joined the faculty at the Medical College of Wisconsin and could not afford the trip. When I explained, they offered to pay my way. I agreed to come assuming that they wanted me to participate in a discussion of my paper.

The food and beverage industry, I would later learn, had established ILSI to study nutrition and food safety. The woman who had called to invite me to the meeting was an employee of Coca-Cola, a major supporter of ILSI. When I arrived at the meeting, I noticed immediately that I did not have a place on the panel of speakers dealing with chlorination by-products. I concluded they must have flown me to Washington to comment from the audience on issues raised by the speakers.

On the second day of the meeting, I found a seat in the vast, generic meeting hall, which was filled with hundreds of engineers, microbiologists, and toxicologists. I listened to one speaker after another, waiting for my chance to talk. Several speakers commented on my study, but none of them offered any specific critique. The first speaker to directly address the epidemiology of disinfection by-products was Gunther Craun, a former EPA employee who had gone on to become a successful consultant to the drinking water industry. He offered his own review of the epidemiological studies on chlorination by-products, concluding that they were inconclusive with respect to the risk of cancer. In his closing comments, he dismissed my study as an inappropriate use of meta-analysis. When Craun finished speaking, I stepped to the microphone reserved for questions.

"You have criticized the use of a meta-analysis in this context. A meta-analysis combines the results of studies using a strict, objective, quantitative scheme to draw conclusions from the literature. A review of the literature, the only alternative, relies on an unspecified, subjective, qualitative scheme to draw conclusions. Is a qualitative, subjective method really superior to a quantitative, objective method?"

Craun simple restated his belief that the meta-analysis should not have been attempted. I started to ask if he considered it appropriate to reject meta-analysis in favor of a literature review, which offers no protection against the biases of the reviewer. As I did the panel moderator cut me off. There was, he said, no time for more questions.

If I had any doubt about the agenda of the EPA, it disappeared as I listened to a speaker on the next panel, Pat Murphy, an epidemiologist for the drinking water office of the EPA. Murphy spent almost her entire time at the podium on an extended critique of the meta-analysis. Murphy, it turned out, had played a key role in an internal EPA review that had evaluated the potential for some sort of meta-analysis of the studies on chlorination by-products. She had concluded that a meta-analysis was not feasible. "The Morris meta-analysis," she explained, "combines apples and oranges and comes up with a fruit salad." She smiled at her own joke.

Had Murphy been speaking to an audience of epidemiologists, the flaws in her critique would have been obvious. However, I believe I was the only epidemiologist in the audience and I know I was the only physician. My experience with Gunther Craun had made it clear that I would have no opportunity to respond in any meaningful way to her speech. It seemed that I had been invited to be seen, not heard.

I was getting my first glimpse of the strange interface between the EPA and the drinking water industry, and I was beginning to realize just how many toes I had stepped on. Without intending to I had indicted the EPA drinking water research laboratory, which had signed off on the safety of our water supply under the existing standards. With a meta-analysis that had taken six months to prepare and write into a manuscript, I had also stolen the thunder from epidemiologists, including those at the NCI, who had spent years on the painstaking process of collecting the data from people all around the country to look for the connection between drinking water and cancer. I also learned that I had contradicted the conclusions of the world's experts on chlorination by-products and cancer who had recently met at the International Agency for Research on Cancer (IARC) and concluded that there was not enough data to draw any conclusions about this relationship. They refused to even allow the possibility that the by-products were carcinogens based on the available data.

Above all, limitations on the use of chlorine could create a massive problem for the drinking water industry. Drinking water treatment is unlike other regulated industries such as the automotive industry, the power industry, or the chemical industry. Those industries are paid to produce products that are not directly related to the quality of the air or water. Pollution occurs as an unintended consequence of the manufacture of cars, power, and petrochemicals. Protecting the environment will not in general improve the quality of their product. Hence those industries have a relationship with regulators that is inherently adversarial. The drinking water industry, on the other hand, is in the business of cleaning water and selling it. Requiring them to produce cleaner water seems consistent with their mission. On the surface it appears that the EPA and the drinking water industry have the same agenda. But the story is more complex.

Drinking water treatment plant managers have two major goals. First and foremost they must prevent massive outbreaks of waterborne disease. This requires disinfecting the water, which usually depends on the use of chlorine. The second requirement is to keep their bosses happy. In most cases that boss is an elected official who is never happy with the thought of asking for more money from his constituents. Using less chlorine, adding systems to remove its by-products, or introducing new treatment technologies each threaten one of these core missions. Furthermore, removing by-products would yield improvements in water quality that their customers, the people paying for the change, might not even be able to detect. In other words the drinking water industry had a tremendous vested interest in my being wrong.

But I soon learned that the industry had other, more urgent problems. As I sat at the ILSI meeting waiting to hear from the epidemiologists, I listened to engineers and microbiologists describing the world of water. What I learned suggested that water suppliers faced a problem that could prove far more destructive than any possible cancer risk from chlorination by-products.

The threat took shape through the words of Joan Rose, a tall, blond microbiologist from the University of South Florida. Dr. Rose described the watery world of protozoa in stark terms. These microscopic creatures lie somewhere between bacteria and true animals. Rose showed data demonstrating that protozoa routinely contaminated rivers and streams throughout the United States. She also showed that specific protozoa often made their way through water treatment plants and into America's drinking water.

As she spoke I thought back to my first encounter with a protozoal infection. As a medical student, I had helped care for a Milwaukee policeman. He lay in a hospital bed as the parasite slowly consumed him. I might have expected him to recover from the infection, but for a blood transfusion he had received two years earlier. The blood contained HIV (human immunodeficiency virus). As a result he lacked an effective immune system and we had no effective antibiotic for treating his infection. So we could only offer him temporary support in a losing battle for his life.

The title of the ILSI meeting, "Safety of Water Disinfection: Balancing Chemical and Microbial Risk," told the story. After decades of believing that water treatment had made major outbreaks of waterborne disease a thing of the past, the industry faced a growing crisis. On one hand mounting evidence suggested that chlorine, an essential ingredient in that success, posed serious, unforeseen risks. On the other work by Rose and other scientists raised the possibility that the corpse of waterborne disease was not as dead as they had hoped.

10

SPRING IN MILWAUKEE

The winds of chance blow through the careers of every scientist, but few feel the effects of those winds more profoundly than epidemiologists. Proximity to disaster can define a reputation. Late in the spring of 1993, as my plane rose up over the blue collar landscape of Milwaukee's south side, chance was on the move. I had been on the faculty of the Medical College of Wisconsin for two years. My father was dying of prostate cancer and I was flying east to be with him and my family. I stared down as the plane veered out over Lake Michigan. The water, normally a cold, dark blue was pale and brown. My thoughts were ahead, in Connecticut with my father. But my future was below me in those turbid waters.

March, as always, was a messy time in Milwaukee. The Bucks were limping toward the end of a dismal season, hopes were riding high on the Brewers with the approach of opening day, and the weather was toying with the city's emotions, swinging wildly between spring and miserable. First came the snow. The second week of the month had dumped six inches on the city before clamping it in a vise of dry arctic air. Nighttime temperatures dipped to three degrees. For almost a week, the snow gleamed under an icy sun. Then warmer air arrived laden with moisture. Alternating waves of rain and snow left the city awash in slush. Snowmelt filled the storm sewers. The rivers grew fat and angry. The Menomonee and the Kinnickinnic roared into the Milwaukee, which sent a plume of brown water surging into Lake Michigan. The rains and runoff of March carried more than mud to Milwaukee. They brought the seeds of catastrophe.

Tuesday, March 30, was a brilliant, sunny day with a gentle breeze. It seemed, for a moment, that winter had left the house, but the next day, it stormed back in like an angry drunk. The wind spun to the west and cold air roared off the plains. The skies opened and temperatures plummeted. What began as heavy rain turned to sleet. As April began, ice glazed the city and a bitter wind blew snow into every corner.

For Mark Rahn, a forty-four-year-old car salesman from Chicago, the April Fool's Day snowfall was little more than scenery. As he walked slowly out of the Milwaukee County Medical Complex, he had other things on his mind. Mark Rahn was trying not to die.

Locked in a pitched battle with leukemia, he had come to Milwaukee to undergo an extreme form of treatment known as a bone marrow transplant. Drawn by the reputation of physicians at the Medical College of Wisconsin, he had subjected himself to radiation and chemotherapy intended to destroy his bone marrow. Then, with his own ability to generate blood cells gone, he received a small amount of healthy bone marrow cells from a donor. He would then need to wait for those cells to find their way to the hollows left by the destruction of his own marrow, set up house, and create new blood cells for him.

Half the battle in surviving leukemia is surviving the treatment, particularly when the treatment is a bone marrow transplant. The drugs and radiation had ravaged the healthy cells in his body. They had eroded the lining of his gut, leaving him vomiting and often unable to eat. As his red blood cells had died off, he had become anemic and needed transfusions to survive. Worst of all the elimination of his white blood cells had stripped him of his ability to fight off infections. A few fungal spores drifting on an air current, an invisible clump of viral particles carried in on a visitor's hand, or even the bacteria from his own gut could kill him.

After weeks in an isolation unit breathing microfiltered air, things were looking up. The chemotherapy had stolen his thick head of hair, the steroids had bloated his normally slender frame, but Rahn was hopeful. There was no sign of the leukemia in his blood. It seemed that the worst was behind him.

On April 3, his first morning out of the hospital, he awoke to bright sun streaming through the window. He had every reason to feel optimistic, but as the day wore on, the feeling was shattered. Painful cramps racked his body and he began to vomit. He searched for an explanation. Perhaps it was from the drugs he was taking to help him survive the bone marrow transplant. Then the diarrhea hit him. Wave after wave. This was not going away. He would soon find himself back on the wards of Milwaukee County Hospital, a return that was both physically and mentally excruciating. Although he had no way to know it, Mark Rahn was not suffering alone.

In the nearby suburb of Wauwatosa, Tom Taft and his wife had already been ill for several days. They had spent much of the preceding week suffering from intense cramps and profuse diarrhea. Most people would have simply gritted their teeth and waited for the illness to subside. Tom Taft was different—as an infectious-disease physician he wanted to know what had hit him. And how it had found him.

High on Dr. Taft's list of possible sources was a dinner the couple had attended at Milwaukee's Italian-American Community Center. Community gatherings and picnics are notorious for their capacity to cause small outbreaks.

Dr. Taft then discovered that his case was far from isolated. Over the course of the week that he was ill, he had begun to receive calls from colleagues at West Allis Hospital asking him to consult on patients who were hospitalized with severe diarrhea. In each case the standard tests had failed to find the organism responsible. The fact that doctors had no lead on the pathogen meant they could only guess at how to treat the disease. Antibiotics did nothing. Their best bet was to help their patients find a way to replace the fluids they were losing because of the intense diarrhea. All over Milwaukee, doctors offered their patients the same advice. Drink water. Drink plenty of water.

By Sunday the hospital was flooded with patients. As he consulted on some of the most severe cases, Tom Taft sensed that he might be seeing the tip of an epidemiological iceberg. For each patient he was seeing, how many more were out there? How many were going to other hospitals or being treated by their doctors and sent home? How many were just starting to get sick? How many would start to get sick tomorrow?

It was the weekend. Snow, ice, and slush covered the sidewalks and streets. When Liz Zelazek went out that Saturday, she assumed that the weather was the only thing keeping people off the streets. Then, late Sunday, she turned on the news. It had been a slow news day, but a story carried by one channel caught her attention. Several Milwaukee drugstores had been stripped of their supplies of Imodium and Kaopectate, over-the-counter drugs used to treat diarrhea.

Later that night her sister called. When Zelazek mentioned the news, her sister expanded on the story. She had just been to the drugstore and the shelves that normally carried medication for diarrhea were empty.

Zelazek made a mental note to herself to mention this report when she arrived at work the next day. As the director of public health nursing for the city, finding the reasons for this apparent run on medication might even become her first task for the next day.

Fifty thousand years ago, vast glaciers, some more than a mile thick, crawled across North America, tearing away pieces of the land as they came. Then, twenty thousand years ago, the ice began to melt. Like geological toddlers, when the glaciers retreated from southern

Wisconsin, they left behind a messy jumble of earth and stone now known as the Kettle Moraine because of its pockmarked blend of hills and hollows. Today the farm town of Eden, Wisconsin, sits in the Kettle Moraine, seventy miles northwest of Milwaukee. Just to the east, in a small valley, a river begins.

The river once wandered through virgin forests of beech and basswood, past dense stands of sugar maples and under the shadows of ancient oaks. The Potowatomi came to eat the fish that thrived in its cool waters and hid in the hollows of its many tributaries. Over time the forests fell to the axes of farmers who had come to brave the wild western frontier of a young America. First the French, then the British, then the Germans all came to stake their claim, strip away the trees, plant crops, and raise livestock.

As immigrants came they built the farm towns that hang on the river like so many beads on a string. From Eden, the river winds south and east past Campbellsport, Kewaskum, and West Bend, past Newburg, Waubeka, and Fredonia. It scrapes along a limestone bluff, just a few miles from Lake Michigan, through Saukville and Grafton, Cedarburg and Thiensville. Then the river finds its outlet. At a break in the limestone, the river makes a final turn to the east and empties into the vast fresh water sea of Lake Michigan. That estuary and the bluff above it formed a favorite spot for the natives who called it simply *milwaukee*, the gathering place by the river.

The river itself also came to be known as the Milwaukee. Just before its mouth, two tributaries add their waters, the Menomonee from the west and the Kinnickinnic from the south. All three rivers slow and widen as they approach the lake, and as they do, the rich topsoil suspended in their current drops to the bottom. Dense fields of wild rice once rose in this lush estuary.

As the city of Milwaukee grew, the river that had attracted people to its banks as a source of food and water grew foul and lifeless. By the twentieth century, the river had become an industrial shadow of its former self. Today the estuary has long since disappeared beneath layers of fill. A lattice of train tracks runs where wild rice once grew.

The Menomonee River, together with the tracks that follow its course from west to east, splits Milwaukee in two. When Polish immigrants arrived in the city in the early twentieth century, they settled the low-lying, marshy area south of the river. On a high bluff facing Lake Michigan, they built St. Stanislaus Catholic Church, the first Polish church in any American city.

In time new waves of immigrants brought new faces, new names, and new cultures to the pews of St. Stanislaus and built other, smaller Catholic churches in its shadow. First Serbs, then the Irish and Italians, and later an influx of Hispanics filled the homes and factories of Milwaukee's south side. Their children came to learn at the district's parochial schools. On that first Monday morning of April 1993, these schools were among the first to register the effects of the outbreak. One of them, St. Adalbert's Parish School, announced that it would not open. The school could not find enough healthy teachers to run its classes.

In nearby West Allis, school administrators spent the morning on the phones scrambling to find substitutes as teachers called in sick. They soon discovered that the substitutes were sick too. They then made plans to combine classes in order to accommodate all their students. Oversize classes would be better than no classes at all. They need not have worried about class size; when the school did open, a quarter of the students never arrived. They were home sick.

On the other side of the Menomonee River, Paul Nannis was already at his desk in a spartan office on the first floor of the city's municipal building. He had risen early that morning and eaten breakfast as his two cats prowled among his jungle of houseplants like leopards in the rain forest. As always he had spent part of the weekend in his office, catching up on work in the unruffled quiet. Like most of his senior staff at the Milwaukee Health Department, he had not seen the news the night before and had no idea of the events unfolding around the city as he headed out into the cool, gray morning. By seven-thirty he was hard at work as Milwaukee's health commissioner.

In a cramped lab two floors above, Gerry Sedmak practiced the art and science of growing viruses. Most bacteria are easy to grow, as they require little more than a dish full of food and a warm place to reproduce. Viruses, on the other hand, are far more finicky. To grow them Sedmak needed a cell that the virus liked to infect and a way to grow those cells. As they grew he would add a sample from a patient and, if the virus were present, it would infect the cells. As he arrived on Monday morning and began to check on the various cultures running in his lab, the phone rang.

Whenever a new bug seemed to be making the rounds, Milwaukee's medical reporters knew to suspect a virus. One of them had picked up on the television news story and wondered if something was brewing. As the city's virologist and an affable source for the press, Sedmak was the go-to guy for viruses. He usually knew what viruses were working their way around Milwaukee, but nothing growing in his lab could explain the run on Imodium. Before long, a second reporter called. Dr. Sedmak didn't have a ready answer. As soon as he hung up, he called Paul Nannis.

Nannis didn't like to see information flowing uphill. He did not want to learn about a possible outbreak from the news media. Whatever was happening, news needed to start flowing downhill and fast.

Nannis could not even begin to unravel this mystery on his own. The first nonphysician to run the city's health department, he had been chosen for the job because of his skills as a manager and communicator. His job was to know how and when to bring the resources of the department to bear on a problem. Nannis began typing a note to Steve Gradus, his laboratory director.

Nannis relied heavily on Gradus, but the two men inhabited separate hemispheres in the world of public health practice. Nannis worked in the land of education, communication, politics, and public relations. Gradus operated, for the most part, out of the public eye, in a world of science, not convention; a world of logic, not nuance. A self-effacing, highly skilled microbiologist, Steve Gradus loved the quiet refuge of his laboratory.

Gradus was looking through his e-mail from behind a thick brown beard and a pair of horn-rimmed glasses when he got the note from Nannis. As he read it, he was not sure what to make of it. Perhaps this was much ado about nothing, a twenty-four-hour virus making the rounds of a neighborhood. On the other hand, it could be the opening salvo of a significant outbreak. The challenge of cornering an unknown pathogen amid the glassware, reagents, and microscopes kept him coming to work each day. But it would take some time before he could begin to hunt for the cause in his own laboratory. For the moment his most important laboratory instrument would be a telephone.

He needed to collect his own data. If a few dozen people were staying home with diarrhea, he had something mild on his hands. If they were going to the hospital, he had a more dangerous pathogen to worry about. The directors of the microbiology laboratories at hospitals around the city would tell him whether or not this was a significant outbreak, and if so help him determine its size and location. His first call made it clear that this was real and serious. The lab at St. Luke's Hospital, one of the city's largest, had been so overwhelmed with requests to test stool samples that they had run out of culture medium.

If a single hospital was seeing a large number of cases of debilitating diarrhea, Gradus's public health training and experience told him that a local restaurant or a picnic or banquet might well have been the source the disease. Spoiled or contaminated food frequently causes sudden, localized outbreaks. With his second phone call, however, the picture changed.

The second lab director told Steve Gradus the same story as the first. The weekend had brought a run on stool cultures. With at least two hospitals involved, the possibility that a single restaurant had caused the outbreak faded. Perhaps a chain of restaurants or a major food supplier was distributing tainted food. Maybe spectators at a Bucks game or the Frozen Four (the final rounds of college ice hockey's national championship, which had recently been played in Milwaukee) had been hit by undercooked bratwurst.

Gradus needed to consider other ways by which this disease might be spreading. It could have passed from person to person. An unwashed hand, a dirty doorknob, or a shared sip of soda could all transmit diarrhea. However Gradus knew that diseases spread in this way tend to cause slow-moving outbreaks. It seemed, for the moment, that this one had spread too far, too fast.

Another alternative, an airborne virus, could cause a sudden, widespread outbreak. Milwaukee had seen that happen a few years earlier when inadequate vaccination levels had allowed a measles outbreak to take hold. But airborne spread requires something to launch the virus. Only after a cough or a sneeze sends millions of viral particles hurtling through a room full of potential victims can a virus touch off such an explosion. In other words, if the outbreak were airborne, Gradus would see respiratory symptoms among at least some of the victims.

To narrow the list of possibilities, Gradus needed more data. He needed to know who was getting sick and what kind of symptoms they were experiencing. He continued calling other labs around the city and began calling emergency departments as well. As they did so, the outbreak began to tell its own story. Few of the patients had coughs, so this did not appear to be airborne. It also did not seem to spread by personal contact.

As any parent knows, children are the perfect vector for the spread of disease from person to person. If the disease was occurring predominantly in children, the outbreak could be spreading through the schools. As the two microbiologists called around the city, however, they learned that Children's Hospital had not been particularly hard hit. Their other calls suggested rather that adult males were showing up at hospitals in large numbers, hardly the demographic of a pre-school outbreak. If the disease was striking disproportionately at adult men, a large public event like a sporting event could be its source. Gradus and the city epidemiologist would need to investigate recent events to look for a source.

Kathy Blair (then Fessler) had not started out to become an epidemi-ologist. She began her career as a nurse in the neonatal intensive-care unit, but the long late hours became more than she could handle. She joined the health department as a public health nurse with the expec-tation that she would have a more manageable schedule. She hadn't counted on the month of grueling days that lay ahead.

Blair had almost no formal training in epidemiology; she had learned on the job by assisting Tom Schlenker, the city's medical offi-cer. As small outbreaks hit Milwaukee, she had worked with Schlenker on the search for their cause. She learned quickly, and when, in 1988, Nannis decided that the city needed an epidemiologist, Schlenker appointed Blair to the job, a decision that was about to be put to the test. Blair would have primary responsibility for the epidemiological analysis of this outbreak.

By the time Liz Zelazek poked her head in the door that morn-ing to tell her former nursing colleague about the report she had seen on the evening news the night before, Blair had already gotten word of the outbreak from Paul Nannis. Her first reaction was to try to find its boundaries. As Steve Gradus called labs and emergency rooms, she called an epidemiologist at the Wisconsin Department of Public Health. She described the cases that were occurring in Milwaukee and asked if anyone else in the state was seeing anything similar. Milwaukee, she learned, was alone.

As news of the outbreak filtered into the health department that Monday morning, Gradus and Blair met with Schlenker to assess the situation. Initial reports suggested that the outbreak had begun five to seven days earlier and had grown slowly before exploding over the weekend. They still had limited information about the number of cases, but during the weekend a single, small emergency room at St. Francis Hospital had seen 145 patients weak and severely dehydrated. Worst of all it appeared that the outbreak might still be growing.

If the disease was still spreading, their task took on a new urgency. Every moment that passed until they identified the source meant more cases. They knew that many of those cases were severe, even life-threatening. Any lost hours could mean lost lives.

Finding the pathogen would take them a long way toward solving the puzzle, but the hospital labs hadn't given them much to work with. They had all tested for bacteria using bacterial cultures similar to those developed by Robert Koch more than a century before. Those tests take forty-eight hours to produce definitive results, but preliminary results were negative. The standard O&P (ova and parasites), a microscopic search for protozoa and parasitic worms had been negative. Viruses can also cause waterborne disease, but physicians did not routinely order tests for them since the tests were cumbersome and effective treatment did not exist. The few viral tests that had been ordered were also negative.

If these results were correct, the hospital labs had either missed something or simply failed to turn over the right stones. Steve Gradus made plans to get samples from hospitals around the city to run his own tests. Blair, working with Gradus, Schlenker, and the state epidemiologist, laid out a plan to begin gathering the epidemiological data that they would need in order to isolate the source of the outbreak.

As they advanced their epidemiological dragnet, one explanation could not be ignored. Even though it seemed unthinkable for an American city in 1993, they had to consider the water supply.

At the time of its creation in 1874, drinking water quality had been a core responsibility of the Milwaukee Health Department. The first laboratory tested only two things: water and milk. In 119 years much had changed. Water quality was now the province of engineers and chemists in the Department of Public Works (DPW). Even though the DPW occupied offices just one floor above the public health laboratories in the municipal building, the two departments had nothing that could be called a working relationship. Except for following up on the occasional call from citizens concerned about their water, the health department had had almost nothing to do with the water supply.

The DPW maintained its own labs for testing water. They no longer relied on the health department, but they did have a responsibility to notify the health department if the drinking water had violated federal standards. There had been no such notice. Assuming there were no problems in the lab and regulations had been followed, a waterborne outbreak seemed unlikely. Nonetheless they needed to be sure.

Milwaukee draws its water from the vast reservoir that is Lake Michigan. Its waters were once so pure a fish could be seen as much as eighteen feet below the surface. Although it no longer had such crystalline purity, the lake was (and is) far cleaner than many other big city water supplies. Two water treatment plants supply the vast network of water pipes that runs beneath the city streets. The Linnwood Treatment plant, a grand, New Deal–era public works project, sends water into those pipes from the north. Water from the Howard Avenue plant, a more recent and less inspired facility, fills the network of pipes from the south. The two supplies meet and mix somewhere in the vicinity of the Menomonee River.

The call from Kathy Blair and Steve Gradus requesting water quality data from both plants left the staff at each puzzled. Why would anyone want data when the water was meeting standards? This was such an unusual request that there was some confusion as to how to provide the data. The staff at the Howard Avenue plant, however, was more than confused; they were reluctant to provide any data.

Blair and Gradus persisted. Their query led them up the chain of command until they spoke with Jim Wagner, the manager of the city's two water treatment plants. Like everyone else they talked to, Wagner told them to forget about drinking water in their search for a cause. An engineer from the old school, Wagner seemed to see their request as a meddlesome annoyance. All measures of water quality, he assured them, had been well within the EPA's standards. Getting the data would require that he slog through handwritten logbooks kept at each plant. Nonetheless, he promised he would get the water-quality data to them the following morning.

By Monday afternoon local news reporters had caught the scent of the developing story. One newspaper reporter asked Kathy Blair about the rumor that this was a waterborne outbreak. Based on what she had heard from the staff at the treatment plants, she assured him that the water supply could not in any way be linked to the outbreak.

As the Milwaukee media prepared to file their first stories on the outbreak, three pieces of the puzzle had already fallen into the lap of Dr. Jeff Davis, Chief Medical Officer at the Wisconsin State Department of Public Health in Madison. Everything about Davis's background, it seemed, had prepared him for this moment. Not only was he a specialist in infectious diseases, but he had also trained as an epidemiologist at the federal Centers for Disease Control and Prevention (CDC) in Atlanta. For fifteen years he had directed the investigation of every significant outbreak of infectious disease in Wisconsin as the state's top epidemiologist for communicable diseases. Now he had a mystery to solve in his home town.

Davis knew that Milwaukee was experiencing an outbreak of gastroenteritis large enough to strip local pharmacies of over-the-counter medications and to close schools and businesses. He also knew that the disease could be severe. It had sent so many people to the hospital that at least one lab had run out of supplies for bacterial cultures. Finally he knew that the standard laboratory studies had come up empty.

So Davis knew he was looking at a massive explosion of gastro-intestinal disease. The pathogen, whatever it was, could cause serious, even life-threatening, illness and did not appear to be bacterial. He sorted through the same set of alternatives as the public health workers in Milwaukee. The pathogen could be spreading by any of four pathways: personal contact, food, air, or water. If it was spreading from person to person, it was more virulent than anything he had ever seen before. Even a fast-moving virus creates a series of small explosions as it hits one classroom or workplace after another. This was a nuclear blast. Not only had it had grown too fast to be spread by personal contact, it had also grown too big to be a typical foodborne outbreak. Airborne outbreaks could be large and spread rapidly, but there was little sign of respiratory disease among those in the hospital. Waterborne outbreaks always involved a major breakdown in the water treatment process. The water in Milwaukee had not violated federal standards and, as far as he knew, nothing was wrong with the plant. Furthermore no major American city had seen a significant outbreak of waterborne disease since the widespread adoption of routine modern filtration and chlorination systems more than seventy years earlier. Nothing seemed to fit.

Nothing, that is, if the infectious agent had come from the universe of the familiar. Davis's years of experience with infectious diseases had taught him all too well that pathogens have a remarkable ability to change their stripes. This could be a familiar pathogen acting in an unfamiliar way. More alarming was the possibility that this was a new or emerging human pathogen, an agent that had never before shown itself or only recently shown itself to be capable of causing disease in humans. As he considered these disturbing scenarios, a third possibility dawned on him. Perhaps this was a somewhat unfamiliar pathogen acting in a somewhat unfamiliar way. If so Jeff Davis had a suspect.

11

THE HIDDEN SEED

By the time Paul Nannis, Milwaukee's Health Commissioner, picked up the morning paper on Tuesday, April 6, the outbreak lay sprawled across the front page. Four more Catholic schools were closed on Milwaukee's south side. At the Milwaukee County Medical Complex, physicians who only saw sporadic cases of severe vomiting or diarrhea on a typical day now found that half their patients had some sort of gastrointestinal illness.

One of those patients was Mark Rahn, the car salesman from Chicago. Just three days after emerging from the hospital having survived a bone marrow transplant, Rahn could no longer eat. The mere scent of food made him vomit. He couldn't even hold down a glass of water. At the same time, the intense diarrhea was draining twelve quarts of fluid a day from his body. Doctors were pouring saline into his veins in an effort to keep ahead of dehydration. The saline, however, was not enough.. Unable to absorb food, he began to decline. To stave off malnutrition, his doctors would soon need to insert a tube into the large veins in his chest and begin total parenteral nutrition (TPN). For the next ten weeks, they would pump food into his blood as his crippled immune system struggled with the invader. The fight against cancer is not a single battle, but a protracted war. He had been ambushed and was fighting for his life.

As Mark Rahn lay in isolation at Milwaukee County Hospital,

Steve Gradus, the director of the lab at the Milwaukee Health Department, sat at his desk reviewing the water-quality records from the water treatment plants. Most of the data looked unremarkable. The records showed that there had been bacteria in the water from Lake Michigan, but that was to be expected. Both plants had added plenty of chlorine. There had even been bacteria in the treated water, but the levels had been low and well within federal limits. In fact, the plant had met all water quality standards throughout the period leading up to the outbreak. Then something caught his eye.

Gradus went down to the first floor to show the data to Kathy Blair, the city epidemiologist. As they looked them over, the phone rang. It was Jeff Davis, the state infectious disease epidemiologist. When Davis learned what Gradus had found, he told him that he would arrive from Madison the following morning. He would come with a team of scientists from the State Department of Public Health. Davis was growing convinced that this was a waterborne outbreak and began to believe he knew the cause.

What Gradus saw in the data was the trace of a series of events that had their beginnings weeks before as Wisconsin thawed and water filled its every pore. In the spring of 1993, like so many springs before it, rain and snowmelt had seeped from the soggy farms, forests, and towns along the rivers that feed Lake Michigan. An ideal solvent drawn inexorably toward the sea, the cold runoff rinsed the landscape of southeastern Wisconsin. By the time the water found its course, it swirled with everything from road salt and rotting leaves to pesticides and cow manure. As the Milwaukee surged toward the lake, fourteen sewage treatment plants and almost a thousand industries poured their wastewater into the river.

The waters of Lake Michigan had spent the bitter cold winter locked in layers. Particles that fell through the icy depths had spent the season trapped on the lake bottom. The slow spring warming altered the density of the layers of water and stirred this elaborate limnological cocktail. The annual mixing sent clouds of detritus surging toward the surface for the first time since the fall. As the river, swollen with snowmelt, crashed into the lake, the water boiled with nutrients, mud, and microbes.

The churning water reached out into the lake farther and farther until it found the mouth of a massive pipe. That pipe stretched back to the shore, crawling along the lake bottom for more than a mile before rising from the water, piercing a high bluff, and tunneling beneath the suburban streets to its origin in the pump room of the Howard Avenue Water Treatment Plant. The lake water that surrounded the jaws of the pipe swarmed with an infinity of particles. The huge bank of pumps with the capacity to pull more than 100 million gallons of water into the plant each day sucked up these particles like a massive vacuum cleaner.

On that day the chemist at the Howard Avenue Treatment Plant was peering into six glass jars of frigid Lake Michigan water. Spring runoff loaded with particles was nothing new for him. He had seen it happen every spring. Controlling this turbidity had always been a routine matter. Unfortunately nothing was routine for Milwaukee's water in the early spring of 1993.

The cloudiness of the water flowing from the plant to the faucets and bubblers of Milwaukee had jumped sharply during the day. The chemist expected to see high turbidity in the water entering the plant in the spring. Even the turbidity of the treated water might be higher than usual, but when he had measured the turbidity of the water leaving the plant earlier that day, it was more than twenty times the level he had measured just one week earlier. The plant operators, it seemed, had a problem.

One of the jars on the lab bench should hold the answer to their problem. The chemist had added a different amount of coagulant to each of the six jars. The chemical coagulant would cause fine particles to clump together so that the sand filters could remove them.

After the particles had settled, he picked the jar with the water that appeared clearest and recorded the concentration of coagulant in that jar. He then walked downstairs to the operations room of the plant and wrote a chemical change order in the logbook. Soon the chemical technician would check the logbook and adjust the chemical feed pumps so that the concentration of coagulant in the plant matched the coagulant in the jar.

The adjustments to the coagulant dose usually brought the turbidity down. To be sure the chemist took a water sample and placed it in the laboratory's turbidimeter. When he read the meter, he was surprised to discover that the turbidity had not dropped. It had gone up. It would continue to rise over the next forty-eight hours and would stay elevated for five days. This was the spike that Steve Gradus would see as the outbreak took shape.

During those five days, the manager of the water treatment plants, James Wagner, carefully followed the data coming out of the plant's laboratory. In addition to turbidity, laboratory technicians tested samples for the presence of bacteria and for levels of chlorine many times each day. The turbidity, to his mind, was higher than he would have liked, but below federal limits. More important the treated water had tested negative for bacteria. This indicated that so few bacteria had survived the treatment process that there was little chance they could cause any illness. The plant, like almost every other treatment plant in the country, relied on chlorine to inactivate any pathogens that might have squeezed through its filters. Wagner made certain they were adding plenty of chlorine to the water.

Wagner still had reason to believe the water was safe. It had not violated federal standards. But something bothered him. As the turbidity continued to defy their efforts to fully control it, he placed a call to Paul Biedrzycki, head of the environmental health division at the health department.

"Where's the flu?" Wagner asked without preamble or explanation. The question landed like a carp falling from a clear blue sky.

"Huh," said Biedrzycki. Not only had Wagner offered no explanation for his question, but influenza normally strikes during the winter, not the spring.

"The flu, what area of town is the flu?" Wagner offered, already at the limit of his willingness to explain the question.

Puzzled, Biedrzycki suggested he call the virologist Gerry Sedmak. Then, as abruptly as he had started the conversation, Wagner hung up. Biedrzycki could only wonder what had prompted the call. He set down the receiver and moved on to the other environmental health issues that concerned Milwaukee. Up to that moment, drinking water had been a minor part of his job. He didn't give the call much more thought until weeks later, when he realized that he had heard the first whisper of an outbreak.

As the mysterious outbreak grew, the laboratory at Milwaukee's health department began to receive stool samples from hospitals around the city. Steve Gradus hoped he would find the key to unraveling the mystery of this outbreak hidden in these samples. Portions would be tested for bacteria and viruses, both in the city lab and at the CDC laboratory in Atlanta. These were the most likely suspects, but Gradus had to consider all the possibilities.

Protozoa, intestinal worms, and other parasites could also infect the human gut. Many of these agents could be found with nothing more than a microscope and a keen eye. Dr. Ajaib Singh and his staff in the microbiology laboratory had been combing through the samples, but had yet to find a single egg, worm, or oocyst. Grodus had ordered more sensitive tests for protozoa, but those would take time. Since the preliminary results of the bacterial cultures also were negative, Gradus still had nothing solid to work with. He had to assume for the moment that the viral studies, still pending, would reveal the culprit.

In the conference room, the battle lines were drawn. On one

side were the experts and administrators from the DPW. They had two labs devoted to nothing but testing drinking water quality and a fleet of engineers, chemists, and microbiologists trained and experienced in the treatment of drinking water. On top of that, they had the authority of the federal government whose experts who had set the standards for drinking water, standards that had all been followed during the week before the outbreak. If the battle became political, Jim Kaminski, director of the DPW, would take over. In the jungle of politics, Kaminski was an anaconda.

The health department, represented by Paul Nannis, Steve Gradus, Kathy Blair, and Tom Schlenker, had little expertise in drinking water. Even Paul Biedrzycki, the person in the health department assigned to drinking water, had no firsthand experience in the treatment of drinking water or the investigation of a waterborne disease outbreak. Furthermore, he was not at the meeting. Nonetheless, they knew enough to ask the team from the waterworks about the rise in turbidity that Steve Gradus had noticed in their records. In response the team from the waterworks pointed out that there was plenty of chlorine in the water and federal standards had not been violated. Drinking water could not have caused the outbreak.

As Gradus pressed his point, describing a waterborne outbreak that had occurred seven years earlier in Carrollton, Georgia, the meeting began to heat up. Gradus pointed out that a rise in turbidity had preceded that outbreak, but the notion that Milwaukee's massive treatment plants, with their small army of experts, might have the same problems that had plagued some obscure city in Georgia tried the patience of the DPW staff. The water simply could not have caused the outbreak. There had been plenty of chlorine in the water. The bacterial counts had been within federal limits. The turbidity levels not only met standards, they were, in his mind, irrelevant.

Tempers flared in a gathering storm of indignation. From the perspective of the DPW staff, a nurse, a microbiologist, a physician, and a social worker were lecturing the waterworks about drinking water. Not only did none of them have any engineering training or experience, they had probably never set foot inside a treatment plant. Then the storm broke. One DPW senior staff member began to pound on the table as he explained that turbidity had nothing to do with pathogens in the water. Turbidity in the water was simply an aesthetic quality, no more, no less. He continued to pound on the table as he spoke, as if the sheer force of his conviction would make it so.

In the end the staff at the health department, despite their belief that this looked like a waterborne outbreak, could not dismiss the DPW's conclusions. An epidemiological hunch could not trump the expertise and experience of the staff at the water treatment plant. According to federal experts at the EPA, the water had been safe. No one had found a pathogen that could be waterborne. This superficial resemblance to an outbreak in a college town in Georgia was simply not compelling. So that evening when reporters asked Paul Nannis if water could be responsible for the outbreak, he told them that Milwaukee's water is tested daily and "looks fine."

By late in the day on Tuesday, April 6, faxes began to arrive at the health department from hospital laboratories around the city. Final culture results had confirmed preliminary findings. There was no sign of a major bacterial pathogen. If indeed bacteria were not responsible for the outbreak, viruses became prime suspects. It would take several days to get results from the viral cultures now running at the CDC and in the health department's own labs.

The staff at the health department continued to work late into the night on Tuesday, hoping for a break in the case. Until they got a clear finding from the microbiology lab, epidemiology was their best hope for some answers. The scope of the outbreak, however, was beyond anything they had ever attempted to investigate.

The next morning brought a growing sense of urgency together with some much needed help. At nine o'clock Jeff Davis and a team of epidemiologists from the State Department of Public Health arrived to join in the hunt. Phones and computers were brought in to a second floor conference room that would come to be known as the war room. Working closely with the health department staff, Davis and his team began to map out a series of studies that would help to define what was happening in Milwaukee.

With thick brown hair, a serious air, and the mustache that, in 1993, communicated a lingering connection to the sixties, Davis had a depth of experience and training concerning infectious disease outbreaks that far outstripped anyone else in the room. He began, however, with the basics. Every epidemiologist is taught to begin an outbreak investigation with the four Ws. What is the disease? Who is being afflicted? Where do they live, work, and go to school? When did they fall ill?

Under Davis's guidance, the team quickly assembled a series of questionnaires and in short order began to make phone calls. They would call hospitals, emergency rooms, businesses, schools, and nursing homes. They would even call the hockey teams that had played in the NCAA tournament. Above all they would call people in Milwaukee. They would talk to those who had been ill, but also to those who had remained healthy. At the core of epidemiology is the search for differences between those who contract a disease and those who do not. One of those differences is the cause.

Gathering all this information was a tremendous undertaking. Liz Zelazek and her staff of public health nurses were brought in to begin making the hundreds of phone calls needed to conduct these studies and to respond to the growing flood of calls coming into the department from an increasingly confused and concerned public. The next three weeks would exhaust every resource and every person the health department could muster.

As Wednesday marched by, the war room buzzed with activity

related to the epidemiological studies. Samples continued to pour into the health department lab from around the city. The Internet was a rudimentary shadow of its current self, so reports of illness relied on faxes and even the fax machines were slow and unreliable. The flood of material and information left the health department staff reeling.

For three weeks no one ate a meal on time. Even when they did find time to eat, they were usually interrupted. On that Wednesday afternoon, three days into the investigation, lunchtime had long since come and gone when Steve Gradus finally sat down to eat. Just as he did so, his phone rang.

When he picked it up, Tom Taft was on the line from West Allis Hospital. "Hey, Steve," he said, "I think I might have something here."

If the battle against this outbreak had become a war, Dr. Taft had become one of the commanders in the ground campaign. He had been called to see more and more patients who were suffering from the mysterious infection. Taft had enough experience that he could often identify an organism just by the pattern of signs and symptoms in his patients. How severe was the diarrhea? How intense were the cramps? How long did they last? Even the appearance of the stools could reveal the culprit. But so far the pathogen had not shown its hand. Then as he stood waiting for an elevator, another physician pulled him aside.

Taft's colleague was caring for a woman with severe diarrhea in the hospital's intensive care unit. A gastroenterologist had already examined her. Using a flexible fiber-optic cable known as an endoscope, he had examined her esophagus, stomach, and small intestine. He saw no obvious explanation for her condition, so he had used a special attachment on the endoscope to clip a few small samples of tissue and had sent them to the pathology laboratory. When the pathologist had examined the samples, he had been surprised. In the lining of her small intestine, he had found what looked like cryptosporidium.

A finding of cryptosporidium by itself would not have been

unusual. It often appeared as a cause of severe diarrhea in people with AIDS. Even the fact that it had put this woman in the ICU was not extraordinary except for one thing. She did not have AIDS.

As soon as he could, Taft made his way to the ICU to see the woman himself. In any hospital the intensive care unit defines a special territory where death holds watch as modern medicine wields all of its technical muscle in defiance. Like all physicians, he had learned to approach the ICU with a calm remove that allowed him to focus on the intellectual challenge of saving lives.

As he entered the woman's room, a pale yellow line traced the echo of her rapid heartbeat on a monitor above her bed. He gazed down at her frail body and examined her sunken features. A pump droned in the background as it steadily pushed saline through a catheter and into the veins of her withered arm. That pump was keeping her alive. He lifted her hand and took her twisted fingers in his own. Years of the slow and ultimately losing struggle against rheumatoid arthritis had contorted them at a tortured angle. Her skin had turned thin and paperlike from the side effects of the drugs used to treat her arthritis, drugs that also suppressed her immune system.

The fact that this woman was overrun with cryptosporidium seemed unusual. The drugs she was taking for arthritis had weakened her immune system, but not to the extent that AIDS would. Nonetheless, Taft knew that strange things happen in medicine. He was considering the possibility that this was nothing more than an interesting case when his pager went off. He recognized the number as the microbiology laboratory.

In twenty years as a laboratory technician at West Allis Hospital, Sandy Schroederus had seen nothing that compared to the first week in April 1993. Never had she seen the volume of samples received by the lab outstrip not only their stock of culture media, but the reserves of the surrounding hospitals from which they could have otherwise borrowed supplies. For almost a week, she and her colleagues in the lab had carefully spread tiny bits of stool samples across whatever culture media they could muster and placed it in the lab's incubator, where the climate controls were tuned to match the moist, junglelike interior of the human body. Again and again the microbiologists at West Allis returned to the incubator, looking for bacteria that could explain what was happening to their town. Each time they opened the door, the warm air hit them in a rush filled with a fertile, musty smell as billions of bacteria divided over and over. Each time they had closed it without an answer.

When Schroederus arrived for work on Wednesday, April 6, she had been hunting for the cause of the outbreak for almost a week. On that day she was due for a break from running stool cultures. It was her day to examine specimens for ova and parasites. Perhaps this was the day she would find something.

Success and survival in a microbiology laboratory rely on a meticulous sense of organization that borders on the compulsive. A momentary lapse can cause a lab worker to misidentify an organism. A mistake in handling a sample from a seriously ill patient can even kill. When a disease has defied efforts to identify its cause, any sample must be handled with extra caution. In the laboratory at West Allis Hospital, the possibility that this was something new and possibly extremely dangerous hung in the air.

Schroederus took several new samples and placed them under the hood, a laboratory bench surrounded by a vented enclosure to help prevent the exposure of the laboratory staff to pathogens. She carefully prepared the slides, placing a drop of iodine on each one. Then, one by one, she scanned the slides under a powerful microscope.

The iodine would stain the parasites and their ova a dark purple. As she went through one slide and then another, nothing jumped out at her. Then one slide struck her as strange. As she scanned through the nondescript jumble of organisms and iodine, she saw nothing that looked like a pathogen, but she had the sense that there was something else there. Something she couldn't see.

The particles and organisms she could see seemed too spread out, as if something were pushing them apart. As she stared into the microscope, the visible defined the shape of the invisible. It appeared to her that the area of the slide that seemed empty at first glance was in fact filled with tiny spheres. She wasn't sure what she had, but the stakes were high. She couldn't afford to miss anything.

So she went on a hunch. She carefully prepared a slide from the sample and added a preparation of fluorescent antibodies. When she looked at the sample, she was stunned. Hundreds of brilliant spheres jammed the field.

She refused to believe her own eyes. She had to be sure there was no mistake. She made a new slide, double-checking every step. When she looked again, any doubt disappeared. The sample was filled with the egglike oocysts of cryptosporidium.

When she spoke to her supervisor, he recalled the biopsy sample from the woman with diarrhea. That had looked like the active form of cryptosporidium but there are no antibodies to the organism, only to the oocysts, so they had no definitive proof. Now they had a confirmed case.

Could this be the organism they had been scrambling to find? To get the input of a physician, they paged Tom Taft.

Taft answered the page and, as he learned what they had found, the pieces of a huge puzzle began to tumble into place. The finding from the woman in the ICU raised his suspicions, but here was an immunocompetent patient with severe cryptosporidiosis. The implications were staggering.

Taft believed he might have the missing piece for which the health department had been searching. He immediately called Steve Gradus to tell him what he had found. Although it would take many more tests to be sure, these two cases raised the possibility that Milwaukee was experiencing an outbreak of cryptosporidiosis among people with normal immune systems. If so, how had so many people become so sick so fast? The answer began years earlier with a meeting between two scientists in the Arizona desert.

The career of a successful scientist may be, as Edison suggests, 99 percent perspiration, but without those rare moments of inspiration, all that sweat is for naught. Seven years before the outbreak in Milwaukee, Joan Rose, a young postdoctoral fellow at the University of Arizona, had such a moment as she sat in the office of Chuck Sterling, a new faculty member in the Department of Microbiology.

Tall and blond, with a blazing smile and boundless energy, Rose shatters any stereotypes of a cloistered laboratory scientist. Raised in the blast furnace of California's Mojave Desert, she had been set on the path of science at a young age. An interest that began with ant farms and chemistry sets gave way to thoughts of medical school and volunteer work as a candy striper at the local hospital. Nothing in the hospital captured her imagination as much as the dazzling array of equipment she saw in its microbiology lab.

By 1980 she had completed bachelor's and master's degrees in microbiology and was starting work on her PhD at the University of Arizona under the tutelage of Chuck Gerba, an iconoclastic environmental microbiologist with a particular interest in waterborne pathogens.

At first glance the deserts of the Southwest might seem like an odd place to go to study water, but no area of the United States cares about it more. In the early 1980s, Arizona's exploding sunbelt population was sucking water from the state's aquifers so fast that they were losing 810 billion gallons of water each year. Today the desperate response to this unsustainable water consumption is a 336-mile canal that cuts across the entire state of Arizona. With a dozen tunnels and siphons and sixteen pumping stations, which along the way lift more than 300 billion gallons of water more than half a mile in the air every year, the aqueduct carries water from the Colorado River to the desert cities of Arizona ending in Tucson at the far corner of the state.

Even with the canal, the state continues to deplete its precious groundwater. A plan is already in place for Tucson to treat and then drink its own sewage sometime in the next ten years. The state's ever-expanding thirst created a critical job for Gerba and his doctoral student. The wastewater the people of Tucson were preparing to drink was loaded with viruses far smaller than the bacteria that had challenged Robert Koch and Louis Pasteur. Rose began working with Gerba to study the occurrence of viruses in treated sewage.

In 1985 Rose found herself with a newly minted PhD in environmental microbiology, two young children to feed, and an uncertain future. She decided to stay on at the University of Arizona for a postdoctoral fellowship. Postdocs are academic purgatory. In a vicious job market, they buy the budding scientist some time to prove that she can get the grants and produce the papers essential to survival in academia.

Her research moved from viruses to a protozoan by the name of giardia. Like bacteria, protozoa are composed of a single cell, but a closer look shows that these cells include many of the key advances in cellular organization exhibited by animal cells and necessary to form higher organisms. More important for Rose and Gerba, protozoa cause some of the most serious human diseases, including malaria and amebic dysentery. Giardia had been recently recognized as a cause of severe, prolonged diarrhea and had been associated with outbreaks of waterborne disease.

Rose had already begun her research on giardia when she met Chuck Sterling. Sterling had started out studying malaria, but he had moved on to look at a closely related protozoan called cryptosporidium. First identified in 1907, cryptosporidium commanded little interest until the 1960s, when veterinarians discovered that it was a major pathogen in livestock, particularly calves. The medical community took notice in 1981, when it began to show up in AIDS patients as a cause of severe, debilitating, often fatal diarrhea.

Rose had never heard of cryptosporidium, but as Sterling laid out the life cycle of the parasite, she was struck by his description of its oocyst, a hard egglike structure that contains the organism in the dormant phase of its life cycle. Chitin, the protein that forms the oocyst, is the same compound that forms the outer shell of ants and other insects. Once in the environment the toughness of the chitin shell allows cryptosporidium to remain viable for months.

As she listened, Rose had a moment of inspiration that would change her career and with it the world of drinking water. This new bug sounded to her like an armor-plated version of giardia. Cryptosporidium, she realized, walked, swam, and quacked like a waterborne pathogen. What's more, Sterling had just developed a tool that would be essential to any further research.

If Rose wanted to study cryptosporidium, she had to be able to find it. Even under a microscope, oocysts are, for all practical purposes, invisible. Not only are they minute, they are also transparent. Its discoverer immortalized this elusiveness with a name that translates as "hidden seed." Before Sterling introduced his improved methods, most oocysts escaped detection. He had shifted the balance in this war of concealment by using a recently developed microbiological tool known as an ELISA (enzyme linked immunosorbent assay). The use of an ELISA required a supply of antibodies to cryptosporidium oocysts and Sterling had just developed the capacity to produce those oocysts in his laboratory.

Sterling's antibodies, like most antibodies, had a shape similar to a pair of pliers. The jaws of those pliers have a unique contour that matches molecular projections that occur only on the surface of their target, in this case cryptosporidium oocysts. An oocyst, when combined with these molecular pliers, would emerge bristling with antibodies.

Sterling's antibodies were not just any antibodies. When exposed to ultraviolet light, the back end of the handles of the molecular pliers glowed like stars. These were the same antibodies that Sandy Schroederus would later use to find cryptosporidium oocysts in stool specimens in Milwaukee. This type of antibody is known as an immunofluorescent antibody and armed with it Rose could test a slide on which she had collected what might be oocysts. If Rose wanted to use Sterling's technique to look for oocysts in water, she had only one problem: getting cryptosporidium onto a slide.

Stool specimens, such as those tested by Schroederus, could contain billions of oocysts, so getting enough oocysts to ensure the test would work was not a concern. But in drinking water, Rose would usually be looking for fewer than one hundred oocysts in an entire liter. Perhaps fewer than ten oocysts. Drinking water containing just a few oocysts in a liter could trigger a massive outbreak of disease. Finding an oocyst in a child's glass of water is comparable to finding a single pea in a glass more than a mile high and half a mile across. Rose's experience with giardia told her that she would need to filter hundreds of gallons of water if she wanted to capture enough oocysts to make the antibodies useful.

Rose set to work in the lab, and by 1986 had a test that, although it could not find every oocyst, could find enough for her to begin looking. She immediately set to work using it to test for cryptosporidium in water supplies. As she collected samples from different locations in the Southwest, she began to find cryptosporidium. Everywhere.

Rose's hunch had proven correct. Cryptosporidium could be found in water, but no one was sure what it meant. She was about to find out. In January 1987 Rose was in the midst of her hunt for oocysts in water supplies when she got a phone call from the EPA.

Earlier that month a physician at the clinic for West Georgia College in Carrollton had been inundated with students complaining of diarrhea. When state and federal epidemiologists came to investigate, they discovered that the outbreak had hit not just the college, but the entire city as well. The outbreak was so widespread that the epidemiologists who came to investigate soon began to suspect the water supply.

The majority of Carrollton's homes still relied on wells rather than the public water supply. As they tried to follow the outbreak back to its source, epidemiologists found that people who had fallen ill were far more likely to have used the public water supply, either at home or at work. They also found that the outbreak had hit nursing homes using the public water supply far harder than those that relied on wells. Using techniques almost unchanged since John Snow first applied them in London, they pointed to the water supply as the source of the outbreak.

Then investigators discovered cryptosporidium in samples from several hospital patients. A massive outbreak of cryptosporidiosis among people with normal immune systems was unprecedented. Since Rose's work with Gerba and Sterling demonstrating the ubiquity of the organism in surface water was still unpublished, the possibility that it was waterborne came as a shock to public health workers. At this point the EPA became involved, including a scientist by the name of Walt Jakubowski, the man who had given Joan Rose her first grant for the study of cryptosporidium. When Jakubowski learned about the outbreak, she immediately came to mind.

When she got the call from Jakubowski, Rose packed up her equipment and flew to Georgia. When she arrived in Carrollton, she collected a thousand gallons of treated drinking water, enough to fill twenty bathtubs to the brim, and ran it through a filter. After the long, painstaking process of isolating the oocysts from the filter, Rose removed a tiny BB-size pellet from her centrifuge.

She spread the pellet across a slide and added a drop of Sterling's antibodies. Then she put the slide under a microscope and flipped on the ultraviolet light. As she scanned it, she began to see small glowing spheres. Cryptosporidium had found its way into the water supply.

When the Carrollton outbreak was reported in the *New England Journal of Medicine*, it seemed like a curiosity. Waterborne outbreaks were assumed to happen in small towns when wellheads leaked or when inexperienced operators mismanaged antiquated treatment facilities. Most of the medical community paid it little attention. Even five years later, when I listened to Joan Rose describe the data that she and others had accumulated showing that cryptosporidium could be found in treated drinking water all over the country, few in the medical community listened and no one fully grasped its urgency.

Jeff Davis and Steve Gradus had read about the Carrollton outbreak and were aware that cryptosporidium could cause waterborne disease. So when Steve Gradus got the phone call from Tom Taft, he immediately recognized its implications. Although he lacked conclusive evidence, it now appeared possible that Milwaukee's water supply had been and might still be contaminated.

The city now faced a high-stakes decision. If in fact the water supply had been compromised, failure to shut it down could lead to hundreds of cases of serious illness. If it had not, the city and its businesses would incur millions of dollars in unnecessary costs, the people's faith in the water supply could be dramatically damaged, and the reputation of the city as well as the seventy-seven food and beverage companies around Milwaukee that relied on that water, including its famous beer industry, would receive a big black eye. For the moment the decision hinged on a single confirmed test from a single patient.

Any pronouncement about the water supply was not for Gradus to make. That burden would fall on Paul Nannis and the mayor. His job was to make their deliberations easier. He had to find more data, and fast. He called the city's hospitals to tell them to begin testing all samples for cryptosporidium, but those tests would take time. In the meantime he called Nannis to let him know about Tom Taft's phone call.

By four o'clock Paul Nannis was standing before the city's press corps. Short with sandy hair and a closely cropped mustache, he did not have a commanding stature, but his friendly manner and ease in public allowed him to convey the difficult news. With his senior staff by his side, he announced that cryptosporidium had been detected and that this finding raised the possibility that Milwaukee was in the midst of a major outbreak of waterborne disease. The press pounced. What had been a local story about a curious disease outbreak had just become national news. For a moment the story teetered on a single laboratory finding, but the plot was about to thicken.

As the press digested the implications of Nannis's message, a member of the laboratory staff rushed down the hall outside the conference room. He whispered his findings to Kathy Blair and Steve Gradus. The results from three of the samples that Gradus had sent for special stains had come back. All three were positive for cryptosporidium oocysts.

That evening, as Liz Zelazek sat in a Milwaukee restaurant and fingered the water glass the waiter had set before her, she hesitated and, for the first time in her life, wondered if she should drink it. She and many of the nurses that worked under her at the health department would be working far into the night to find out if the water was safe. She had taken a break from that search to meet several friends for dinner. As she looked around the restaurant she saw a sea of water glasses and realized that the danger they might hold was invisible to everyone in the room but her. Then, not quite prepared to accept the possibility that something she had always seen as healthy and refreshing could be the source of disease, she took a sip.

When Zelazek left the municipal building earlier that evening, she knew that all the senior staff working on the outbreak would be meeting with the mayor to discuss the implications of finding cryptosporidium. As she walked back up Wisconsin Avenue to return to work, she saw that the lights in the conference room were still blazing.

She knew in that moment that the crisis had reached a turning point. The meeting with the mayor had gone on far too long.

In the conference room, John Norquist, the city's lanky, curly-haired mayor was presiding over a debate that would decide the fate of the city's water supply. Five more patients at city hospitals had tested positive for cryptosporidium. Nannis and his staff from the health department wanted action, while Kaminski and his staff continued to argue against declaring the city's water unsafe. Doing so would have huge costs in terms of both finance and public confidence. Even with the new findings, the total number of cases was still only eight, and they had no direct evidence that cryptosporidium was in the water. Kaminski always seemed to have secret levers he could pull to control the political machine, but this time none of them was working.

Kaminski's position crumbled when the mayor noticed that Jeff Davis, the state's infectious disease epidemiologist, was drinking a Coke. The mayor asked Davis if he would drink the water. When Davis said no, the decision was made.

Norquist's staff quickly called a press conference and at nine o'clock he announced to the city's reporters and photographers that he was recommending that all residents of the greater Milwaukee area boil their drinking water until further notice. In doing so, he admitted that there was still some uncertainty about the cause of the outbreak. "We are erring on the side of caution. We do not know enough of where it came from. We are not even sure it's from the water supply."

In New Haven, my father was dying. As tumor cells dug into his bones, the hospice nurse increased his morphine dose in a losing battle against the excruciating pain. The whirlwind of memory, emotion, and family relations had put me in an unintended news blackout for almost a week. So when I called a friend at home on the night after Norquist's news conference, I was unprepared for the news from Milwaukee. "Haven't you heard? We're national news," she said.

"We're boiling our drinking water. I feel like I'm living in a developing country."

The premonition I had felt while listening to Joan Rose came rushing back to me. I had known it would happen, but I had never imagined it would find its way into my water pipes. As one of just a handful of people in the country who specialized in drinking water epidemiology, this was the opportunity of a lifetime and chance had just dropped it on my doorstep. But the same roll of the dice had put me a thousand miles away.

The next day, Good Friday, my father succumbed to his cancer at his home in New Haven. That same day, in Milwaukee, people began to die from bad water. The first victim was a cancer patient whose chemotherapy had left her vulnerable to the disease. Cryptosporidium overwhelmed her, used her to create trillions of oocysts, and left her to die from dehydration. Mark Rahn, the forty-four-year-old leukemia patient, would spend ten weeks in intensive care before winning his fight with cryptosporidium but many others were less fortunate. The deaths that followed included the very young, the very old, and people with suppressed immune systems either from medication such as chemotherapy or from diseases, particularly AIDS.

Also on Good Friday, five days after the initial recognition of the outbreak, the Howard Avenue Treatment Plant shut down its pumps. There was, however, still no guarantee that the water had caused the outbreak. No one had found the organism in the water and cryptosporidium had other ways of spreading.

Within a few days, analyses had found small numbers of cryptosporidium oocysts in the water supply, but it didn't seem like there were enough to cause the kind of outbreak that Milwaukee was experiencing. The task of finding the "hidden seed" fell to Paul Biedrzycki, head of the Division of Environmental Health. With limited training in waterborne pathogens, working on his first waterborne outbreak, he now had to perform a magic trick. He needed to go back in time almost two weeks and collect several hundred gallons of water for testing.

Biedrzycki needed to find someone who had collected huge samples of Milwaukee's drinking water every day for the past several weeks and stored them. The water treatment plants and many of the city's industries such as the breweries collected and sampled water, but the samples were small and were usually discarded after analysis. The "smoking gun" for the outbreak, it seemed, had long ago been flushed down Milwaukee's drains and sewers.

As the hope of finding the instrument of disaster dwindled, a health inspector on Biedrzycki's staff had an idea that would lead him to it. The search had failed because it had been directed at the wrong target. The "smoking gun" was not lying on the bottom of a vat of aging water. It was frozen in a block of ice.

Biedrzycki rushed to the vast freezer owned by Milwaukee's largest commercial ice supplier. As he entered the cavernous building, a snake of frigid air crawled into his lungs. It seemed as if they had uncovered winter's lair. Once inside he gazed at rack after rack filled with massive, glasslike blocks of ice. Each four-hundred-pound block marked a moment, literally frozen in time.

As the plant manager guided him through the crystal maze, he peered into the massive blocks of ice, staring through to the clear center as workers at the plant raised them from the vats in which they were stored. The tour took them back into the past until they reached the blocks frozen a week before the outbreak began. As he stood there, Biedrzycki felt sure he had found what he had come for. Looking into them, he could see that the heart of each block was dark, almost black with fine particles.

Whatever had made the drinking water cloudy was locked in the ice. To go back in time, all Biedrzycki had to do was melt the ice. The blocks were too big to move, so he set up a pump and filter at the plant. As the ice blocks slowly melted, the particles began to float freely in the water just as they had weeks before when they flowed from the faucets and water fountains of Milwaukee. He filtered the water and sent the specimen to a lab in Illinois, one of only three in the country that could find cryptosporidium in water. When the results came back, the answer was clear. Each block contained thousands of cryptosporidium oocysts.

On April 19, the last piece of the puzzle fell into place as the results emerged from the core epidemiological survey of the outbreak. The health department staff had initially believed that the outbreak had attacked a few hundred people. As the outbreak wore on, they began to think that it might have hit as many as a thousand. No one fully grasped the power of water to spread disease. When Jeff Davis completed his analysis of the data, he estimated that 400,000 people had fallen ill. Of those, 4,000 had been hospitalized and, according to initial estimates, more than 100 had died. Some had avoided the disease either by drinking little or no tap water from the Howard Avenue plant, or by having some immunity from a previous bout of cryptosporidiosis. One thing, however, was clear. Milwaukee had just experienced the largest outbreak of waterborne disease in the history of the United States. A jagged mountain of illness had crashed to the surface, but how much of the iceberg remained hidden across the United States remained to be seen.

Carrollton Water Treatment Plant

Collecting water in Mali

PART III

At War with the Invisible

> Water, water, everywhere,
> And all the boards did shrink;
> Water, water, everywhere,
> Nor any drop to drink.
>
> —SAMUEL T. COLERIDGE

12

DRINKING
THE MISSISSIPPI

Washington rushes to correct environmental problems with all the speed and agility of a glacier on quaaludes. During the 1980s the legislative glacier ground to a halt as the Reagan administration systematically eliminated funding for drinking water research and the EPA refused to issue new regulations. In 1986 Congress ordered the EPA to issue an array of new tap water regulations, but the EPA simply did nothing until an alliance of citizens' groups took them to court. The agency finally buckled under a court order and issued new rules in the early nineties. In 1992 the arrival of a new administration brought hope that the process would again move forward, but treatment plants are expensive, and the drinking water industry resisted. Then came Milwaukee.

In many ways the evisceration of the EPA budget for drinking water research and the subsequent failure to correct that imbalance made the disaster in Milwaukee possible, even inevitable. That outbreak could have been predicted and should have been prevented. Several studies, not only from Joan Rose and others in academia, but from the industry's own experts, had shown that cryptosporidium oocysts routinely occurred in surface water throughout the United States and were frequently present in treated drinking water. The outbreak in Carrollton served notice that a massive outbreak in a treated water supply was far more than a theoretical possibility. Despite all those warnings, neither the EPA nor the industry sounded the alarm. That silence made it possible for the water supply of a major city to operate as if cryptosporidium did not exist.

After Milwaukee, the drinking water industry could no longer ignore the threat. Or so it seemed.

In the fall of 1996, I got a call from Erik Olson, an attorney with the Natural Resources Defense Council (NRDC). As the head of the drinking water program at NRDC, Olson had become a persistent, effective, and often lonely voice for tighter federal rules for drinking-water quality. This made him the official thorn in the side of the drinking-water industry. Even after Milwaukee, it had taken three years of wrangling before the EPA was ready to bring stakeholders together to talk about new rules for drinking water. Olson wanted me to come to their next meeting.

I had made many trips to Washington to help Olson push for tighter regulations for safe drinking water. He and I had often shared our frustration at the inaction of the federal government. The stakeholders' meeting seemed to offer our best hope of progress, but there was sure to be a fight. Nothing could make me miss this showdown.

A few days later, on a cool, gray November morning, I met with Olson at a coffee shop in Georgetown. With an unruly splash of receding light brown hair, wire-rimmed glasses, and a slightly disheveled look that reflects intensity of purpose rather than lack of care, Erik Olson seems an unlikely man to take on an entire industry single-handed, but in many ways he has. A former EPA employee, Olson had joined the mass migration out of the agency that began in response to Reagan's unabashed anti-environmentalism. He began work for the NRDC in 1991.

As I sipped a cup of overcooked coffee, Olson explained the

game. The EPA was prepared to establish new rules to reduce both the levels of disinfection by-products and the risk of waterborne pathogens in drinking water, rules required by the latest revision of the Safe Drinking Water Act. Any rule change would trigger a federal law known as the Federal Advisory Committee Act, which required that the agency seek input from those affected by the rule in the form of a consensus recommendation. This initial meeting would lay the groundwork for the committee and its task. The precise composition of the committee had yet to be determined, but that would happen in the next few weeks. After that, traveling the path to consensus could take months or even years.

I choked down the remnants of an overly sweet bran muffin, finished my coffee, and we crossed the street. I knew my reputation among those in the industry and approached the meeting with some trepidation. The civil engineers who have taken on the relatively unglamorous task of cleaning dirty water spend their careers making sure that water is clean and safe. Coming from the world of academic public health research, I was an outsider. Although I had never intended to do so, my research, particularly the study on chlorination by-products, suggested that they weren't doing their job right. In the same stroke, I had implied that the regulators had allowed this to happen. I had stumbled into a nest of busy bees and had managed not only to infuriate the bees, but to provoke the beekeeper as well.

The room bustled with experts and consultants from every segment of the drinking water industry. The American Water Works Association, the Association of Metropolitan Water Agencies, the National Rural Water Association, and the Chlorine Chemistry Council were all well represented. Senior managers from water utilities all over the country had come to observe. Erik had managed to bring in a few voices from the public health community, but we were clearly outnumbered and absolutely outfunded.

Because of my infamy among the drinking-water industry and

the regulatory community, most of the people in the room seemed to know my name. Few of them, however, knew me by sight, including Lynn Pappas from the EPA drinking water office in Cincinnati. Much of the meeting was devoted to describing EPA's work on drinking water. On the first day of the meeting, Pappas rose to describe the state of the agency's work on epidemiology related to drinking water.

Pappas began by announcing that work by the EPA's Office of Research and Development on "the Morris study" was almost complete. She explained that Charlie Poole, an epidemiologist at Boston University, was repeating my study of chlorination by-products. This was the centerpiece of their work on drinking water epidemiology. In fact it was their only work on drinking-water epidemiology at that time. The final results, she assured the audience, would soon be ready for release.

I sat in the audience, stunned. Ordinarily if a study is of such great interest to government scientists, they talk to its author, possibly invite him to come present the work in more detail and make plans to extend the research. But no one from the EPA had ever contacted me to discuss my original study. Until that moment I did not even know they had planned to reanalyze it. I did know Dr. Poole and knew that he was one of the few environmental epidemiologists who had been critical of the study, but he had no special expertise in drinking water or meta-analysis. I later learned that the initial instructions from the EPA to Dr. Poole specifically instructed him to reanalyze only the papers I had considered in my original meta-analysis despite the fact that several major epidemiological studies of chlorination by-products and cancer had been published since its release that would have provided critical new data. Why was the EPA so interested in rehashing a study that was now six years old rather than moving the science forward?

I don't know all the answers to that question, but many of them lie in the chasm that has opened between public health and drinking-water treatment since the epidemics of cholera and typhoid gave birth to them as twin sciences in the nineteenth century. The

engineers, chemists, and microbiologists who now dominate the province of drinking water dwell in the land of physics and chemistry, a
realm ruled by unwavering principles that obey finely honed equa-
tions. Laboratory science makes good sense to them, but the squishy
conclusions of epidemiology do not sit well, especially when they
imply the need for substantial and expensive changes in the tried and
true methods for treating water.

Three lines of evidence, primarily from epidemiologists, were
driving the move to tighten regulations on drinking water. Research
had continued to suggest that chemicals formed during the chlorination of drinking water increased the risk of cancer (the Poole study
notwithstanding).* A new front had opened when a series of studies
raised the possibility that these chemicals might also cause miscar-
riages, retard the growth of the fetus, and induce birth defects. None
of this research gave clear, irrefutable evidence, but there was enough
smoke in the air to encourage a prudent observer to pull the alarm
or at least make sure nothing was burning. Just as the industry and
its regulators were trying to digest this information, cryptosporidium
hit. Like the massive chest pain that finally gets us to the doctor, the
events in Milwaukee had forced a more thorough examination of our
drinking water. That closer look had begun to reveal evidence of more
widespread disease.

When I began to look at the data that emerged from the Milwaukee
outbreak, the fact that the water had not violated any federal drinking
water standard during the period leading up to the outbreak stopped
me in my tracks. How could the largest recorded outbreak in U.S.
history have been caused by water that was, according to the federal
government, safe to drink? When I attended meetings related to
drinking water, people in the industry were always quick to describe
the Milwaukee outbreak as an isolated episode caused by a rare conver-

---

* At the time of paperback publication, numerous studies, including two independent meta-analyses, have confirmed the fundamental conclusions of my study.

gence of conditions, but I remained unconvinced. If standards were not violated, how could we be sure that this was such an unusual event?

To begin to answer this question, I first requested records of routine tests for water quality from Milwaukee's two treatment plants. The waterworks, more forthcoming in the wake of the outbreak, sent me data for a period of two years, which included the fifteen months leading up to the outbreak. From it I could see that elevated turbidity had not been a rare event in Milwaukee. It had been routine.

If the turbidity of Milwaukee's water had often been elevated in the months and years leading up to the outbreak, it seemed possible, even likely, that many cases of waterborne disease had occurred that were not recognized as such. But could I prove it? There was no way to know if the water had contained cryptosporidium. Water suppliers do not routinely test for specific pathogens. Instead they use surrogates like turbidity and the presence of certain common types of bacteria to test the water. I could not prove that Milwaukee's drinking water had caused cryptosporidiosis before the 1993 outbreak. But I had an idea that might shed some light on the question.

Through the billing office at the Medical College of Wisconsin, I was able to determine the number of adults and children treated for gastrointestinal illness by Medical College physicians on each day of the two-year period. With the help of a talented Russian statistician, I pulled the two data sets together. As we worked, a disturbing picture began to emerge.

Our analyses showed that each time the turbidity of Milwaukee's drinking water rose the number of serious cases of gastrointestinal illness tended to rise about one week later. This time lag is critical. In clinical studies, volunteers exposed to cryptosporidium usually developed symptoms after six to ten days, a time delay known as the incubation period. No other common waterborne agent that causes gastrointestinal infections has such a long incubation period. In other words, the evidence available suggested that Milwaukee's water had been causing cryptosporidiosis for months if not years before the outbreak.

This was not the first evidence to suggest that human pathogens could penetrate the barriers erected by conventional water treatment plants to cause disease without it being recognized as waterborne. Two years before the outbreak in Milwaukee, Pierre Payment, a Canadian microbiologist, studied two groups of randomly selected homes in Montreal. One group received an in-home water filter. The other group served as a control. Both groups received water from the city's state-of-the-art water treatment plant. He followed the health of both groups for twelve months, paying particular attention to any gastrointestinal illness. He found that the water filters reduced the rate of significant gastrointestinal illness by about one-third despite the fact that the city had a water treatment plant that was as good or better than most other cities in North America.

The results of the Milwaukee study and the Canadian study suggested that our water supply was not as safe as we have long assumed and waterborne disease was far more common than anyone had imagined. How much waterborne disease occurs every year in the United States? The CDC routinely publishes reports of all the known outbreaks of waterborne disease in the United States, but these usually list only a few thousand cases. However even those who collect these data concede they are a gross underestimation of the scope of the problem.

These numbers rely on a method known as passive surveillance or what one might call coffee cup epidemiology. Grab a cup, sit down, and wait for a phone call. Nothing happens until a state or local health official concerned that he or she is seeing a waterborne outbreak contacts the CDC. Someone, somewhere, has to start thinking a group of cases is waterborne. Then that person must take the initiative to contact someone in an agency with enough epidemiological firepower to conduct a thorough study of the event. Only if that agency conducts a proper study and can uncover compelling evidence to show that the outbreak is waterborne will the cases find their way into the CDC report on waterborne disease.

This approach might have worked when we were worried about the lingering remnants of typhoid fever, dysentery, and cholera in a country still upgrading its water supply. These diseases rarely go undiagnosed and when found immediately raise suspicions about drinking water. Modern water treatment has done an excellent job of driving these diseases into obscurity in the developed world. But passive surveillance was never intended to catch outbreaks with less obvious links to the water supply. According to Dennis Juarnek, a drinking-water epidemiologist at the CDC, "we came within a whisker" of failing to recognize that the outbreak in Milwaukee was waterborne. Not only were the holes in this system big enough to drive a truck through, they almost allowed the biggest truck in American history to pass by undetected.

So how much waterborne disease is there in the United States? The truth is that we don't know, but several lines of evidence suggest that millions of cases of waterborne disease, perhaps more than ten million, may be occurring every year in the United States. How is this possible? How could water treatment plants that rely on methods refined over the past century allow pathogens to reach our taps? A journey along the path of a single city's water from its source to the tap tells this story best.

The devil, invisible and relentless, lurks in the details. As I prepared to write this book, I tried to identify a water supply that faces the range of challenges confronting water supplies around the world. I picked a city and made plans to visit on August 22, 2005. I didn't give it a second thought when I had to cancel the trip at the last minute. I assumed that I could reschedule for a few weeks later. One week later hurricane Katrina arrived and much of New Orleans, the city I had intended to visit, had disappeared beneath her floodwaters. The disaster has only made the story of that city's water supply more informative.

• • •

Lake Itasca appears unremarkable among the hundreds of lakes that dot the pockmarked flatlands of northern Minnesota. At its north end, it gives rise to a river that appears equally unremarkable except for one thing—its name. There, just south of Bemidji, Lake Itasca officially gives birth to the Mississippi. Headwaters, however, are a fiction of mapmakers and explorers. In reality the Mississippi begins on a thousand hillsides as a thousand streams take form.

Altogether, 65 million people live along the baroque filigree of rivers and streams that form the Mississippi watershed. All of them drink from those streams or the aquifers that feed them. All of them discharge their sewage into that same watershed. Those people along with the industries and farms in the watershed use 135 billion gallons of water each day, the majority for irrigation and agriculture. Sixteen billion of those gallons return directly to the streams and rivers of the watershed as treated and untreated wastewater. The remainder runs off the land, seeps into groundwater, is drawn into crops, or evaporates.

In the northeast corner of that watershed, the upper Allegheny flows out of western New York, swells in the valleys of the Appalachian mountains, and marries the Monongahela in the hills of western Pennsylvania. Pittsburgh presides over the joining of the rivers and marks the birthplace of that new waterway, the Ohio River. Just three miles downstream, when the weather is dry, a steady stream of Pittsburgh's filtered sewage pours into the newly formed Ohio River.

A heavy rain, however, sends Pittsburgh back in time. Like many of the world's largest cities, Pittsburgh's sewage flows into the sewers that drain its streets. When a rush of rainwater overwhelms the treatment plant, the sewers overflow and send a vile soup of raw sewage and runoff from the city streets into the rivers around Pittsburgh at a rate of 16 billion gallons each year.

Pittsburgh is not alone. Since Edwin Chadwick first made the decision to send human waste into the London sewers, cities around the world have followed suit. In the United States, 752 cities have combined sewers (i.e., sewers that carry both street drainage and domestic and industrial wastewater), including 152 in Pennsylvania alone. Sewage treatment plants were an afterthought in the design of these systems. None of them can handle the gushing torrent that fills these sewer pipes during a heavy rainfall. Together they dump 861 billion gallons of untreated sewage into American rivers and streams every year, roughly half of it into the Mississippi watershed.

Wastewater treatment does not remove all hazards from sewage. Pathogens, particularly those that are resistant to chlorine, can and do survive the process. Also the process is not designed to remove chemical contaminants. Some chemicals are chemically altered during treatment or trapped in the sludge removed from the wastewater, but others pass into the river. Household chemicals, the chemicals from industries that send their waste to the treatment plant, and even the drugs that we consume and excrete find their way into treated wastewater. Treatment plants send everything from hormones to antidepressants into our waterways. The chemical stew emerging from treatment plants can be potent enough to change the sex of fish swimming in the river.

Laden with the effluent from Pittsburgh's sewers and factories, the Ohio leaves Pennsylvania and lumbers westward. It flows past Cincinnati, Louisville, and hundreds of smaller towns and cities. The liquid boundary between north and south, it accumulates water and waste on its path to the Mississippi.

Three thousand miles from the roots of the Ohio, in the mountains of southwestern Montana, three trout streams merge to form the Missouri River, at the northwest corner of the Mississippi River watershed. Near its origin, coursing through the eastern slopes of the Montana Rockies, the Missouri is a clear, swift mountain stream. As the land flattens, the river slows and grows into the Big Muddy, waddling across the Great Plains toward St. Louis.

The Missouri and the upper Mississippi pass through some of the richest farmland in the world. As they pass, the chemicals of modern agriculture flow downhill and into their tributaries. From a watershed that includes 65 percent of America's cropland, more than one million pounds of pesticides and the runoff from 6 billion pounds of fertilizer find their way into the Mississippi River and its tributaries each year.

The watershed is also home to the majority of the 60 million pigs, 99 million cows, and 1.3 billion chickens that live in the United States. These animals drink from the watershed, consume its crops, and produce vast quantities of waste. The watershed must absorb more than 120 billion gallons of animal waste every year, far more waste than is produced by the humans with whom they share the basin.

So too thousands of mills and factories ooze poison into the Mississippi River basin. Along its lower reaches in Louisiana, the great river encounters one of the densest concentrations of chemical plants and refineries on the planet. Together, the states that comprise the basin send 120 million pounds of toxic chemicals into their waterways including just under a million pounds of known or suspected carcinogens every year.

As an old professor of mine used to say, "The solution to pollution is dilution." The Mississippi absorbs this immense quantity of pollutants into a far more immense volume of water. Every day 396 billion gallons of water surge through its final reaches. But with this massive load of contaminants, dilution is not enough.

By the time the Mississippi reaches Louisiana, it has become a massive muddy snake of mythic proportions. Then, in the land of the bayous, the river begins to break down. Two hundred miles from the Gulf of Mexico, it splits in two and would redefine its course each flood season were it not for human effort to tie it to the map with hundreds of miles of towering levees. At its south end, the river disintegrates into a twisting web of small rivers before disappearing into the Gulf of Mexico.

The Mississippi arrives at the Gulf of Mexico like an exhausted traveler dropping its luggage, some 440,000 tons of sediment each day, at the door. To say the Mississippi River ends in Louisiana is to mistake the map for reality once again. In truth the Mississippi invented southern Louisiana. The continental shelf ends at Baton Rouge. The rest is sediment left from a time when the Mississippi ran free. Today that sediment is trapped in a river that follows a course defined by man and rides out into the ocean.

Thick with organic matter and fertilizer, the river spills into the Gulf of Mexico so laden with nutrients that it creates an explosion of algae and bacteria. This microscopic frenzy of life and death is so intense that it sucks the oxygen from the water over an area the size of New Jersey. Within that area, known as the Hypoxic Zone, no plant or animal life can survive. No fish, no shrimp, no crabs, nothing.

One hundred miles upstream from the river's mouth, on a crescent of sediment left as the river slows to turn a corner, the French built a city and called it La Nouvelle Orléans. The name has been anglicized and the city has grown, but the Mississippi remains its life force. Today when New Orleans is thirsty, it kneels down by the Mississippi, the drainage ditch for more than 40 percent of the United States, bends low to the water, and drinks.

The motives must have been convincing, indeed, which could induce colonization upon such an uninviting and insanitary waste. A low, flat, marshy area, subject to disastrous inundation at all seasons of the year, pest-ridden, infested with malaria in its most pernicious forms, without a feature to commend it and menaced on every side by seen and unseen enemies—such was the little crescent-shaped village named in honor of the Regent of France, when laid out in the midst of cypress swamps and willow jungles in the year of our Lord, 1690.

—*Martin Behrman, mayor of New Orleans, speaking to the Convention of the League of American Municipalities, September 29, 1914*

In the late nineteenth century, much of New Orleans drank rainwater captured in cypress cisterns. Like the swamps that surrounded the city, the cisterns gave mosquitoes a place to breed, which in turn advanced the spread of malaria and yellow fever. Those who did not have cisterns filled earthenware jugs with water from the Mississippi, allowed the mud to settle, and drank. The water seemed clear enough to drink, but was in reality dirty enough to kill. Typhoid fever and dysentery were endemic and, together with mosquito-borne illnesses, helped make New Orleans one of the most disease-ridden cities in the United States.

In 1891 Col. L. H. Gardner, superintendent of the New Orleans Water Company and a swashbuckling Civil War veteran who had served in the Orleans Light Cavalry, was searching for a source of pure water for the city. The Mississippi offered an endless supply of water, but few people believed it could be purified, not because of bacterial contamination, but because of the vast amounts of sediment it contained, particularly during flood season. Lake Pontchartrain, to the north of the city, seemed a promising alternative, but New Orleans tilts to the north and at the time much of its runoff and sewage flowed into the lake. Efforts to find an adequate supply of clean groundwater beneath the bayous had not borne fruit.

Then Colonel Gardner received a proposal that seemed to give him a way out. The National Water Purifying Company would provide the city with 14 million gallons of drinking water each day. Furthermore, they guaranteed its purity, claiming the water would be "crystal clear water, free from opalescent hue," a coloration apparently related to the oil that seeped from the bayous. To do so they proposed to build the largest mechanical filtration plant yet constructed. Gardner signed the contract sensing that the offer was too good to be true. It was.

Unbeknownst to Colonel Gardner, National Water Purifying had offered to install the plant over the vigorous objections of its largest stockholder, Albert R. Leeds. Leeds was a professor of chemistry at Stevens Institute who held not only several of the key patents for the filter that would be built in New Orleans, but also the first U.S. patent for the chlorination of drinking water. Despite his concerns, by 1893 the company had installed thirty filters, each thirty feet long and eight feet in diameter, in a water treatment plant along the banks of the Mississippi.

The plant was an unmitigated disaster. The muddy water clogged the filters so quickly that they had to be cleaned every four hours. George G. Earl, who became superintendent of the New Orleans Sewerage and Water Board in 1892 after the filters were already under construction, concluded that the water they produced was "not fit for any use except lawn sprinkling." Much of the sediment slipped through the filters, clogged the pipes, and resulted in water flowing from the taps of New Orleans that was "often muddier than the river itself." The New Orleans Water Company, a private company under contract with the city of New Orleans, refused to pay for the filters and National Water Purifying collapsed after losing a court battle to force payment. In this infamous bit of engineering hubris, National Water Purifying failed to grasp a lesson almost as old as Egypt.

Three hundred years earlier, Prospero Alpini, an Italian botanist and physician, had described watching as an Egyptian man filled his camel-skin bag from a cloudy tributary of the Nile and poured it into an oblong clay vessel. The man took a handful of crushed almonds, thrust his arm into the vessel, and stirred it violently. After three hours most of the fine mud had settled and the water was ready to drink.

The Egyptians were not adding almonds for flavor. The oils from the almonds helped suspended particles clump together, a phenomenon known today as coagulation. The practice was so common in Egypt that cakes of almonds were sold in the market just for water purification. By the late nineteenth century, a range of chemical coagulants had been introduced that provided far more effective treatment than crushed almonds. In designing their treatment plant, National Water Purifying had used a coagulant, but tried to save money by not building sedimentation basins, which would have allowed the particles to settle before reaching the filters. Instead dense clumps of Mississippi mud clogged their filters. The disastrous results dissuaded New Orleans from filtering its water for another fifteen years.

Faced with the relentless threats of rampant waterborne and mosquito-borne disease and with no viable alternative to the water from the Mississippi, the city embarked on a bold two-part plan in 1906. The most visible and dramatic element of the plan involved the installation of massive pumps intended to operate almost constantly to drain the swamps around New Orleans. The drainage would surround the city with thousands of acres of new land. The second part of this plan involved the construction of one of the country's largest water treatment plants on seventy of those acres. The city fathers were determined to avoid any repeat of their disastrous first attempt to drink the Mississippi. The Carrollton Water Treatment Plant would use coagulation and sedimentation in conjunction with sand filters, an approach now standard for the treatment of water from an unprotected source.

Water treatment seems simple: filter, disinfect, drink. The devil, invisible and relentless, lurks in the details. Today a cryptosporidium oocyst that has washed off a dairy farm in northern Louisiana and floated down the Mississippi into the intake of the Carrollton plant would face a daunting obstacle course if it hoped to reach a glass of drinking water and infect a new human host—daunting, but not impassable.

The pumping station for the Carrollton Treatment Plant can suck water out of the Mississippi River fast enough to fill a swimming pool in about ten seconds. The water boils with thick clouds of particles, the accumulated debris of the river basin. As soon as the water leaves the pump, coagulant flows into the huge pipe and treatment begins.

Packs of long, snakelike molecules of coagulant wrap themselves around particles, neutralizing their electrical charge. Ordinarily that electrical charge would help to keep particles apart. Now, as the uncharged particles collide, they begin to stick together, like drops of oil in a glass of water. By the time the water reaches the treatment plant, the clumps have grown so large that they begin to sink and form a layer of sludge on the bottom of the plant's sedimentation basins. If our oocyst avoids getting stuck to one of these aggregates, it flows onto the plant's filters.

These beds of sand lie at the heart of the obstacle course. From a human perspective, it is hard to imagine anything significant passing through this thick layer of fine-grained sand, but this perception simply reflects the stubborn myopia of our human-scale world. To help us view this filter with a microbe's eyes, imagine an oocyst enlarged to the size of an average human adult. When we enlarge the filter to match, the smallest grain of sand would grow to the size of a four-story building. The largest would become taller than the Sears Tower.

At this scale one can easily imagine our supersize oocyst tumbling through the gaps between gargantuan grains of sand. Some of the oocysts would get stuck, but many others would find a way through. Bacteria, slightly smaller than an oocyst, could also slip through this maze. For a virus bypassing a filter is even easier. Relative to our human-size oocyst, this protein-encrusted virus would be like a child's marble tumbling through a chaotic stack of boulders the size of apartment buildings.

The coagulant makes the oocyst stickier and makes the filter more effective. As microbes and other debris accumulate in the top layers of sand, they narrow the pathways between sand particles. This accumulation of particles and microorganisms is so essential to the effective operation of a water filter that a clean filter cannot be put online until it forms and must be taken offline before that critical film becomes old and thick and cracks begin to form that would allow pathogens to pass through again. A filter under proper operating conditions will remove ninety-nine out of a hundred oocysts.

One would think that removing 99 percent of the problem would eliminate it, but oocysts can be present in large numbers. If we took the ten thousand oocysts that would fit into the period at the end of this sentence, added them to a liter of water, and poured it over a filter bed, the liter that trickled out would still contain a hundred oocysts. This explains how Rose and others often found cryptosporidium oocysts in treated water.

If we remember that viruses are far smaller and harder to remove than oocysts, it is clear that filters alone provide only limited protection against pathogens. The engineers who designed the treatment process for New Orleans understood this too. Just before filtration they bombard the oocyst with chlorine. A powerful oxidant chlorine burns through the outer skin of pathogens like molecular napalm. The bacteria are usually the first to die. With a tougher shell, viruses last longer. As they can pass through filters more easily, it is critical the operators add enough chlorine to destroy them. But with the arrival of our oocyst, plant operators have a problem.

When microbiologists who study cryptosporidium want to isolate oocysts from a stool sample, they add bleach to it. The extremely high concentration of chlorine in the bleach wipes out most other microbes, but the shell of the oocyst is so tough that the bleach can't cut through it. What remain are viable, infectious oocysts. In other words, there is no way for operators to add enough chlorine to kill cryptosporidium. This explains how cryptosporidium devastated Milwaukee even though plant operators maintained required chlorine levels at all times.

To varying degrees, we all drink from the Mississippi. Water supplies around the country face some if not all of the problems confronted by pre-Katrina New Orleans. Our source waters are universally tainted. In other words, most Americans consume treated sewage.

Almost every water supply in the United States relies on treatment processes similar those used in New Orleans. These filtration systems do not remove 100 percent of pathogens and many of those pathogens are to some degree resistant to chlorine, the chemical used almost exclusively to disinfect drinking water in the United States. To make matters worse, the chlorine used to protect us from waterborne disease may threaten our health in other ways including cancer, still-births, and birth defects.

When the EPA convened the Federal Advisory Committee on Disinfection By-products and Microbial Risk in 1996, these were the problems laid before them. With the events in Milwaukee, it seemed that the writing was not just on the wall, it was written in bold, flashing neon.

The tiny oocyst certainly got the industry's attention. No one wanted a disaster like Milwaukee's to occur on his or her watch. In the drinking water industry, however, big changes mean massive capital projects with long lead times. Over the next nine years, I would watch drinking-water industry lobbyists do whatever they could to slow down and dilute the regulatory efforts that followed.

At one of the committee's regular meetings, I came face-to-face with the industry's resistance to major change. Ed Means, the representative from the American Water Works Association (AWWA), the largest industry trade group, was laying out its position on turbidity limits. Means, smooth and quick-witted with surfer-boy good looks, allowed that the AWWA would accept a reduction in the maximum allowable turbidity for treated water from 5.0 NTU to 2.0 NTU.

I listened in disbelief. The turbidity of the water in Milwaukee during the spring of 1993 never exceeded 1.7 NTU. In other words, the drinking-water industry was proposing a standard that would continue to put a federal stamp of approval on the water responsible for the largest waterborne outbreak in United States history. I had been appointed as a technical adviser to the committee and it seemed that some technical advice was clearly needed. At the next pause in the discussion, I reminded the committee members of the turbidity level in Milwaukee. I assumed they had simply forgotten the exact numbers and would appreciate the correction. I assumed wrong.

At the next break, the professional mediator who was running the meeting pulled me aside. Committee members from the drinking-water industry, she told me, were extremely upset that I had spoken out. Technical advisers, according to strict protocol, should only give information if they are specifically asked to do so by members of the committee. I had served as a technical adviser to many different committees and had never before been chastised for offering advice. Moreover, the previous meetings of this committee had all been relatively informal. My comments, it seemed, had inspired not gratitude, but a new desire for strict adherence to protocol.

The first round of negotiations by the committee stretched on for almost a year as the stakeholders struggled with scientific uncertainty and the inevitable tension between costs and ideal public health protection. The many sides inched their way toward consensus. In large part that consensus was driven by an analysis of what improvements could be made by refining the operation of existing treatment plants rather than introducing any fundamental change in the treatment of drinking water. Even a consensus limited by those constraints proved fragile. At the committee's final meeting, it was almost destroyed.

As the committee members prepared to sign off on a proposed change to the rules for treating drinking water, the moderator who had brokered the deal went around the table asking all the members if they could go back to their organization and sell the deal. From Ed Means of the American Water Works Association to Erik Olson of the NRDC, one committee member after another agreed to the new rules. Just as this began to take on the air of a formality, the question passed to Roy Heald.

Tall and angular, Heald represented the National Rural Water Association. With far shallower pockets than the large cities, the small communities he represented were wary of major rule changes. Nonetheless, he had remained remarkably quiet throughout a year of meetings.

Heald, it seems, subscribed to an extreme economy of words.

Uttering the first words I could remember falling from his lips in almost a year of meetings, he said simply, "No, I can't sign this." Apparently, he had been sitting through the entire meeting for the sole strategic purpose of blocking the process at the point of maximum leverage.

The thud of jaws hitting the floor was almost audible. After months of tough negotiations, the other committee members seemed ready to string Heald up by his heels. The moderator called for a break and she and the EPA administrators huddled with Heald in an effort to assuage his concerns. They brokered a deal that would allow the agreement to survive and the process to proceed. A process that had begun with preliminary discussions in 1992, even before the Milwaukee outbreak, would produce rules that were not scheduled to go into effect until 2002. As the process inched forward, problems continued to plague the drinking-water industry.

13

DEATH IN ONTARIO

rank Koebel just ignored the instructions. A brooding bull of a man with short black hair and the ruddy complexion of an alcoholic, Frank didn't like being told what to do. As foreman for the Public Utility Commission (PUC), he was responsible for maintaining the water supply for Walkerton and the surrounding towns. The chlorinator for well 7, the main source of water for the city, hadn't worked for almost a week. His boss, who happened to be his older brother, Stan, had just left town for a weeklong conference. Before he left, Stan told Frank to install the new chlorinator. Frank decided he had more important things to do.

Installing the new chlorinator just never seemed urgent. After all, it had been sitting in storage for over a year. A few days more or less wouldn't matter. Besides, chlorine didn't seem to do much besides ruin the taste of the "cool, clear, crisp water" that flowed from the aquifers deep beneath southern Ontario. Even when the automatic chlorinator was running, Frank rarely set it to add the required amount. If he did, customers complained. In fact, the wells had run for decades without chlorine until a new regulation required it. When he or Stan was thirsty, they liked to drink the raw, unchlorinated water straight from the well.

To be precise, Frank had already taken the first step in installing the new chlorinator. Two days earlier he and two PUC employees had opened the bypass valves and removed the old chlorinator. Up until that point, the old equipment, while fraught with problems, could still chlorinate the town's water. They left it sitting idle, disconnected, and useless. As they shut the door, the steady whir of the pump meant that well 7 was still running, sending raw water into Walkerton.

On May 8, 2000, three days after Stan Koebel left, rain began to fall. Over the next two days, a steady downpour drenched the town and the farms around it. Along highway 9, the main highway through Walkerton, construction workers slogged through a growing sea of mud. They had exposed the rotten vasculature of pipes and had laid new pipes to replace them. To do that, they would need to cut into the existing water mains to connect the new pipe.

Even Frank Koebel, who had gotten most of his training on the job, understood that opening the pipes in the distribution system posed a risk of contamination, just as a cut in a vein or artery raises the threat of infection. As the spring mud poured into the trenches, the danger increased. If pathogens got into the system, the only thing that would stop them was chlorine.

If the rain worried Frank, it did not convince him to install the chlorinator. It appears that he was not much for vigorous exercise. At forty-one he had already had two heart attacks. The town's two other wells, 5 and 6, both had chlorinators. (Wells 1–4 had been closed long before.) Sometime during the dark rainy night, for reasons that remain unclear, wells 6 and 7 shut down, leaving just five to supply the city.

A small, pious farming community parked halfway between Lake Huron and Lake Erie, Walkerton is the kind of town where nothing ever happens. But Frank Koebel and the storm had just planted a bomb beneath the peace and quiet.

The Koebel brothers' faith in the quality of groundwater was not entirely misplaced. Thick layers of soil, sand, and clay usually filter out most pathogens from water on its way to a deep aquifer. In fact, the beds of sand at the heart of our water purification systems are little more than minuscule imitations of the immense filters that overlie our groundwater. There is evidence that some microbes, particularly viruses, can survive deep below the surface, but they do not appear to pose a major disease risk. The wells in Walkerton, however, were not as safe as the Koebels imagined.

On May 12, in the midst of a ferocious rainstorm, a newborn foal struggled to his feet under the watchful eyes of David Biesenthal, a farmer and veterinarian. After five days of rain, the skies had opened wide, dumping three inches of water onto the fields where newly planted corn was just breaking the surface. Biesenthal had taken advantage of the warm weather just two weeks earlier to sow his fields. The rain, together with the rich manure from his cows and their young calves he had spread on his fields, would give the new crop a great start.

Most of the rain that fell on the Biesenthal farm drained toward Silver Creek. One field, however, sloped gently toward a clump of trees growing on a piece of land owned by the town. David Biesenthal had never noticed the small cinderblock building hidden in those trees. He never heard the pump that ran inside that building. He never realized that the shaft for well 5 reached down into the groundwater just thirty feet below that pump.

Stan and Frank Koebel had water in their blood. They had grown up around the pump houses of the wells of the PUC where their father had risen to the rank of foreman. As soon as Frank finished school, he went to work for the Walkerton PUC. Twenty-five years later, in May 2000, he had his father's old job.

Five years older, with gray hair, a mustache, and a nervous expression, Stan Koebel had always been Frank's boss in one way or another. After thirty years at the PUC, he was now the general manager. In addition to the water, Stan was responsible for Walkerton's electricity. The imminent deregulation of the electrical industry meant that electricity, not drinking water, had occupied most of his time. He tended to leave the operation of the water supply to his younger brother.

Stan returned to the offices of the PUC at six A.M. on Monday morning, May 15. His first task was to deal with an employee at the PUC shop with whom he was having a bitter dispute. Stan planned to have him fired. Before he left for the shop, he looked at the computer control system and noticed that well 7 was not running. Assuming the chlorinator was installed and it had been turned off in error, he turned it on.

Stan drove to the PUC shop where he learned that the chlorinator for well 7 was still sitting in its crate. Moments later Frank walked into the PUC shop.

"Mind if I have the day off, being that I worked Saturday?"* Frank asked.

Frank had spent most of that day repairing flooded electrical equipment at the high school. If it crossed Stan's mind to say anything about the chlorinator, he held his tongue. His bigger, stronger, and smarter younger brother had always chafed under his leadership and Stan knew it.

"Yeah, go ahead," Stan said. The chlorinator could wait another day, but since it was Monday, they would need new water samples. "Just get Al to sample well seven before you leave."

Stan then drove over to highway 9 to check the work on the water mains. He arrived to find the excavation trenches filled with mud and rainwater. He called Al Buckle, a maintenance worker for the PUC, and asked him to bring over some sampling bottles. Once those bottles were full, he packed them in a box and sent them to A&L laboratories with a note. "Please rush. Thanks, Stan," it read.

Four years earlier the provincial leadership had joined the 1990s obsession with privatization and shut down the water testing laboratory at the environmental ministry. Suddenly all water utilities had to find a private laboratory to do their testing. Walkerton had found a lab at the time, but that company had just closed, forcing Stan Koebel to find a new lab. He had just started sending samples to A&L, a Canadian franchise of a large laboratory services company in the United States.

---

* Lines of dialogue are drawn either from reports of government hearings or from Colin Perkel, *Well of Lies* (Toronto: McClelland and Stewart, 2002).

Two days later, on the morning of May 17, Stan Koebel got a call from a supervisor at A&L Laboratories. Tests on the samples from highway 9 had come back positive for coliform bacteria. Water supplies do not test for all bacteria. Instead they use two tests for so-called indicator bacteria. The coliform test is one of the two tests and the positive test simply meant that live bacteria were in the water. It was news of the second test results that got the attention of Stan Koebel. Preliminary results from the second test, which looks for a particular species of bacteria known as *E. coli*, had come back positive. *E. coli* come almost exclusively from human and animal feces.

All water suppliers occasionally find *E. coli* in their water. Often these bacteria are harmless. But they do not normally appear in groundwater. Still Stan Koebel was not overly concerned. He had an abiding faith in the safety of the town's water. By that afternoon, however, more results arrived. Almost all of their samples were positive for coliform and *E coli*. Still worse, the quantitative tests showed that the contamination was massive. Stan Koebel couldn't ignore the results. He decided it was time to install the chlorinator.

Tens of thousands of different strains of *E. coli* inhabit the colons of vertebrates around the world. Although the majority of those strains are harmless, the remainder can cause gastrointestinal disease of varying severity. One of the most dangerous strains is *E. coli* O157:H7, a recently evolved form of the bacterium. The poisons it contains can shred the lining of the gut. Blood then pours into the victim's colon, resulting in the bloody diarrhea that characterizes the disease. Its victim's stools are often nothing but bright red blood. What follows, however, can be far worse.

As *E. coli* O157:H7 makes its way into torn blood vessels, it continues to release its poison. In the bloodstream those toxins tear at the delicate cells that line the blood vessels and shower the body with blood clots. In severe cases it destroys the blood vessels in the kidneys causing them to fail. With no way to purify their blood, these patients begin to poison themselves. The normal products of their own body accumulate to toxic levels and they begin to drown in the by-products of their own existence. Many of these patients die. Those that survive the ordeal of intensive care and dialysis often have chronic medical problems.

In addition to humans, *E. coli* O157:H7 can also infect cows.

On May 18, 2000, Kristen Hallet, one of only two pediatricians working in the region, was called to the emergency room of the Owen Sound Hospital to see a seven-year-old girl from Walkerton. Aleasha Reich had fallen ill the day before and spent the evening drinking fluids as her family doctor recommended. She arrived in the emergency department with intense cramps, vomiting, and bloody diarrhea.

Bloody diarrhea has only a handful of possible causes, none of them common. Dr. Hallet had seen this particular constellation of symptoms only twice before in her career. One was a child she had seen six months earlier who was suffering from an infection with *E. coli* O157:H7. The other, a nine-year-old boy, was lying in a hospital bed just upstairs.

One patient with bloody diarrhea is an unfortunate victim in need of medical care. The simultaneous appearance of two such patients, however, suggests something worse and demands action on a broader scale. Dr. Hallet worried she might be seeing the results of a foodborne outbreak. By the time she called the regional public health official the next morning, all hell was beginning to break loose in Walkerton.

On the afternoon of Friday, May 19, just before the start of the holiday weekend, Dave Patterson, assistant director of regional public health, learned about Dr. Hallet's phone call and asked his staff to investigate. They soon discovered that something was indeed happening in Walkerton. At Mother Teresa School, the town's Catholic elementary school, twenty-five students were home sick. Eight students were sent home from the public school. The disease had also begun to strike the local nursing and retirement homes. Eight people with bloody diarrhea arrived at the emergency room of the small local hospital. A local physician had seen twelve more people with diarrhea. It was Victoria Day weekend with a holiday on Monday, but as Dave Patterson listened, he could see his plans for the long weekend evaporate. He picked up the phone and called Stan Koebel.

The tiny seeds of concern in Stan Koebel's brain began to grow as he answered the call from Dave Patterson. Those seeds had been planted by two earlier calls from people in town worried that the water might have caused the growing outbreak. Just as he had with the other callers, Koebel insisted that the water was fine. Patterson asked if the heavy rain and flooding could have had any effect on the water, and Koebel reassured him that this was unlikely. He never mentioned the problems with the chlorinator or the positive test results.

If there was a problem, Stan Koebel believed he could fix it. He was sure that the construction work was the source of the problem. By pumping chlorine into the system, he hoped to eliminate any contamination. He believed the appearance of chlorine in the water on the far side of the construction would mean the situation was under control. He worked late into the night, scuttling furtively between well 7 and a fire hydrant near Mother Teresa School. He collected water samples and tested them for residual chlorine beneath the light of an oblong moon. Driving home with midnight approaching, he still failed to fully understand that the house of cards he had constructed was collapsing all around him.

That same night Tracey Hammel's heart sank as she opened the diaper of her two-year-old son, Kody. He had been vomiting all day, but she was not prepared for a diaper full of blood. Through a miserable night, she hoped he would improve, but Kody only got worse. By ten o'clock the next morning, he was weak and listless. When she called the emergency room at the Walkerton Hospital, the nurse told her that the hospital was "backed up."

"When can I come?" she asked, unsure how much longer her son could wait. The nurse suggested she come at four that afternoon. In the meantime she advised her to keep Kody hydrated. On the advice of the nurse at the hospital, Tracey Hammel had spent the next two hours forcing water into her son's mouth with a syringe. By noon Kody was limp and his eyes seemed to roll back in his head.

Stan Koebel, meanwhile, had returned to well 7 before sunrise and continued to push chlorine into the system. Most of the passersby who saw the hose connected to the fire hydrant outside Mother Teresa School had no idea what Stan Koebel was doing. Bob McKay, however, knew all to well. He had worked under Stan for the past two years. He knew about the test results. He wanted to make sure someone else knew as well. Later that morning Christopher Johnston, one of the few people working through the holiday weekend at the Ontario Ministry of the Environment, received an anonymous phone call. The water in Walkerton, he learned, seemed to have a problem.

Early Saturday afternoon Christopher Johnston called Stan Koebel. He told Koebel he simply wanted to "find out what's going on."

Koebel struggled for an answer. "We had a fair bit of construction and there is some concern—I'm not sure, we're not finding anything . . . but I am doing this [flushing and chlorinating] as a precaution . . ."

Then Johnston put the question to him directly, "So you haven't had any adverse samples, then?"

Koebel's web had grown tangled, but he hoped he could continue to deceive. "We've had the odd one," he conceded, "you know, we're in the process of changing companies, because the other company, it closed the doors, so we are going through some pains right now to get it going."

As Stan Koebel tried to evade the probing questions, Tracy Hammel held her son, Kody, in her arms, the strength seeping out of his tiny body. She and her husband had finally rushed him to the Walkerton hospital to find it overflowing. The nearest empty hospital bed, they discovered, was in Owen Sound. They raced along the long straight roads that sliced through the farms of southern Ontario. When they finally reached Owen Sound, the hospital laboratory would confirm the presence of E. coli O157:H7.

One of the perverse twists of *E. coli* O157:H7 is its response to antibiotics. As the drugs destroy the bacteria in the bloodstream, they burst open and release the poisons within. This sudden release of toxins dramatically increases the risk of kidney failure. The doctors in Owen Sound knew this and told Mrs. Hammel there was little they could do but give him fluids, watch, and wait. Then Kody's kidneys failed.

The Owen Sound Hospital was not equipped to manage kidney failure. The nearest hospital that could perform dialysis was more than a hundred miles away in London, Ontario. Paramedics rushed Kody to a waiting helicopter.

Time and chance seemed to be conspiring against the tiny boy as the powerful motor of the helicopter spun into action and lifted him into the spring sky. The people of Walkerton who saw the helicopter flash overhead on its way to London had no idea that it carried one of their children. As the week wore on the red and white emergency helicopters continued to come, pounding the air like giant angry insects.

As the helicopters rose, hovered for a moment, and veered to the south again and again, it seemed for a time as if Walkerton were at war. Among their passengers was two-year-old Mary Rose Raymond, the daughter of a physician from a neighboring town who had taken her daughter to Walkerton for a Mother's Day dinner just one week earlier, and Lenore Al, a retired librarian.

In the end Stan Koebel's frantic struggle to erase disaster failed. In a town of just 5,000 people, 2,300 fell ill during the outbreak. Hundreds were hospitalized, many in intensive care. In what his doctors termed a miracle, Kody Hammel survived the weeks of dialysis. Mary Rose Raymond, Lenore Al, and five other victims would die from their infections.

A government inquiry would eventually trace the cause backward from well 5 to the manure on David Biesenthal's farm. The invisible world beneath the surface of the farm did not provide the homogeneous thirty-foot-thick filter that the Koebels had imagined. It now appears that death had come to Walkerton through ribbons of gravel and fractured rock that allowed water and manure to pour into the aquifer beneath the well.

The story of Walkerton might seem to hold few lessons for the future of water. How could the poor judgment of two brothers in a rural town be relevant to the safety of our entire water supply? This tragic tale offers a window into many of the challenges we face as we try to maintain safe drinking water into the future. The problems in Walkerton began with complacency and misplaced confidence in the safety of source water. Those problems were compounded by a failure of water treatment driven, in this case, by a failure of the plant operators. Perhaps most stark is its demonstration of the deadly threat posed by emerging pathogens and the potential for those pathogens to be waterborne. Complacency, the potential for treatment failure, and emerging pathogens pose a threat to water supplies throughout the developed world.

In the United States new rules for drinking water were still on hold as events in Walkerton unfolded. It had taken the EPA five years just to issue rules making the turbidity levels that preceded the Milwaukee outbreak illegal. Until then a utility could have produced water identical to the water that caused the cryptosporidiosis outbreak without violating federal standards.

The final rules, the first rules to actually require communities to test for cryptosporidium oocysts in their water together with more stringent rules for chlorination by-products were due for release in 2002. As 2002 approached, with a new administration in place, the EPA balked. For three years, unseen hands held the final implementation of the new rules in limbo.

In the summer of 2005, Erik Olson from the NRDC filed suit to force the release of the new rules on pathogens and disinfection by-products. The process ground slowly forward. Then, with their release still pending, a new, unprecedented disaster struck at America's drinking water.

14

SURVIVING THE STORM

A storm was coming. The workers at the Carrollton Water Treatment Plant on the west side of New Orleans braced themselves, like an army at the ramparts preparing for a siege. They had packed extra food and clothes and had sent their families to higher ground. They felt sure that they would be safe at the plant. Its thick white walls, hidden steel skeleton, and heavy red tile roofs gave testament to the golden era of civil engineering in which it was built. All the workers at the plant had been through other hurricanes. The plant itself had survived whatever monsters had emerged from the swirling waters of the Gulf of Mexico for almost one hundred years. This storm, however, would defy their collective experience. This storm was called Katrina.

As darkness fell the winds began to rise. Sheets of rain raked across the seventy-acre campus of the plant. For days the storm had sucked energy from the warm waters of the gulf. As she hit land and unleashed that energy, Katrina began to rip the cities of the Gulf Coast to pieces. She peeled off roofs, sucked out windows, and uprooted trees. As power lines snapped and transformers exploded in sparks, the coast of Louisiana and Mississippi plunged into darkness.

In New Orleans the Carrollton Water Treatment Plant seemed prepared. In a city of storms, one could never take electricity for granted. Beneath the two smokestacks that tower over the Carrollton

Water Treatment Plant six boilers power huge generators capable of producing enough energy for a small city. As the power grid went down, the power plant whirred into action. In a darkened city, the water treatment plant didn't miss a beat. But then things got bad.

As the winds rose above a hundred miles an hour, an ornate eight-foot-high window in the power plant exploded, sending glass and bits of the mangled window frame across the floor of the power station. Rain flooded through the opening, soaking the electrical equipment. Within minutes the motor control unit short-circuited and burst into flames. Acrid smoke began to fill the building.

In the control room of the plant, another crisis was unfolding. On a normal day, the huge water pumps at the plant would send 120 million gallons to the city through six massive water mains with the flow rising and falling in response to the demand for water. Just before the storm arrived, with the city half-empty and industries shut down, the demand for water had dropped and the plant was pumping out clean water at a reduced rate of only 50 million gallons per day. But then, as Katrina tore through the city, the demand for water suddenly doubled. The huge pumps whined in a struggle to match the demand. In an empty city, this sudden increase in demand for water could mean only one thing. Somehow Katrina had broken an immense drinking water pipe. With each passing minute, somewhere out in the raging blackness, another thirty thousand gallons of drinking water were flooding from a ruptured pipe and into the streets of New Orleans.

Nine miles to the east, nine miles closer to the ferocious winds that spiraled around her eye, Katrina tore at the sheet metal buildings of the East Bank Sewage Treatment Plant. Built in the 1990s, the plant offered an essential improvement in the treatment of New Orleans sewage, but no one had wanted it for a neighbor. Needed by all and wanted by none, the plant had been banished to a man-made island amid a purgatory of swampland and open water east of the city. Most of the city's wastewater flowed through the grid of treatment basins etched into a rectangle of landfill. The location, chosen for its financial and political expediency, was an invitation to disaster. In the darkness of August 29, the four workers assigned to ride out the storm watched disaster arrive.

As the storm surge began to flood the island and the narrow road connecting it to solid ground, the four men climbed three floors to the control room of the sludge incinerator to escape the rising water. The howling wind strained and tested every seam in the corrugated steel building. Like a ferocious cat toying with a birdcage, Katrina seemed determined to find an opening. Suddenly she ripped out a large window overlooking the plant and reached into the control room with claws of fury.

The four men rushed to the stairwell. Their first thought was to climb higher, away from the rising water. They ran to the forth floor as the storm raged outside. They had escaped Katrina for a moment, but she continued to scrape and claw at the steel skin of the building. Inside, the men could only wait and listen as she searched again for a weak point. Again, she found it. In an instant she took hold of the roof and peeled it back in a horrific explosion of bursting rivets and twisting metal.

As the building disintegrated around them, the men scrambled down the stairs toward the rising water. On the second floor, they found their last, best refuge, a bathroom with concrete walls and no windows. As they huddled there, Katrina's winds seemed to diminish. They began to hope that the worst was over, but the hurricane had other plans. Wind was not her greatest weapon.

Katrina had already sent a massive storm surge toward the city from the south. The waters of the gulf raced up the narrow concrete channel that confines the Mississippi and crashed into the city. At the same time, her winds circled around the back of New Orleans and pushed the waters of Lake Pontchartrain toward New Orleans from the north. Water from both directions roared into the canals that slice through the city and connect the lake with the river. The rising water slammed against the levee walls. The tired ramparts had no chance against the weight and erosive force of the water. It was only a matter of time until it found the weak spots in the walls that protect the city. When it did the catastrophe hit its stride.

As the first gap opened in the Industrial Canal, less than a mile and a half from the wastewater treatment plant, a wall of water crashed into a residential neighborhood and toward the wastewater treatment plant. It flooded the low-lying buildings around the plant within minutes. Water soon rose to the second floor, trapping the four men at the remains of the East Bank Sewage Treatment Plant between the floodwaters and the shattered upper floors.

The disaster was complete. With all the specially designed motors and controllers for the plant under at least ten feet of highly contaminated water, the plant had been dealt a devastating blow. A helicopter would finally reach the stranded crew two days after they took refuge in a bathroom. For the moment they were struggling to survive.

Back across town, the Carrollton Water Treatment Plant seemed to be faring better. A worker had forced his way through ferocious winds and vicious rain to the valve that controlled the ruptured water main. By shutting down a huge portion of the water supply, they could save the rest of the system, just as a tourniquet saves a hemorrhaging patient. The power plant looked like a battlefield with piles of empty fire extinguishers littering the floor and the smell of burned plastic hanging in the air. The ferocious struggle seemed over. For a few moments, the workers at the Carrollton Water Treatment Plant thought they had made it through the hurricane bruised, but not beaten. Then word came that the 17th Avenue levee a few miles to the north had broken. Floodwaters were surging south through the streets of New Orleans. It was only a matter of time until the plant would be underwater.

They had already seen what water could do to their generators. As floodwater ran down Eagle Street toward the center of the plant, they had to make a decision. If they tried to run the system through a flood, they could destroy the generators and any hope of restarting the plant. There was only one choice. So for the first time in ninety-nine years, the plant shut down. They could only hope that the flood would spare the equipment they would need to bring it back on line.

Beyond New Orleans Katrina would shut down 1,200 water treatment plants and 269 sewage treatment plants. In a matter of hours, a single storm crippled sanitation and water supplies across the Gulf Coast, leaving vast areas effectively uninhabitable. Recovery would take weeks, months, or even years. As raw sewage flowed into lakes, rivers, and the gulf, health officials tried to prepare for the threat to the public.

New Orleans, as we have come to know it, is a fantasy, a technological magician's trick. This sleight of hand relies not on smoke and mirrors, but on pumps and levees. The levees lay the foundation for deception. Ignore them and one might think that the Mississippi has always flowed through a fixed and regular channel rather than one that can routinely shift dozens of miles in any direction. One might assume that New Orleans sits on dry ground overlooking the Mississippi rather than ten to twenty feet below it. One might conclude that floods are a disastrous exception rather than an essential component of the river's natural cycles.

As any magician knows, the best place to hide the secret to the trick is out in plain sight. Dozens of massive pumps spread throughout the city run almost constantly and some have been doing so for almost a hundred years. With the capacity to suck a billion gallons of water out of New Orleans every day, they pile illusion upon illusion to make the fantasy almost completely convincing. If one failed to notice them, one might forget that most of what we now call New Orleans was once a deadly malarial swamp. One might not remember that without the pumps a minor storm, a mere inch of rain, would inundate the city with floodwater. Until Katrina the illusion was so compelling that even periodic dire warnings went unheeded.

As New Orleans began to emerge from the floodwater, one of the major concerns that prevented people from returning to the city was the lack of clean water. Even after the city was pumped dry and drinking water began to flow, city residents were warned to boil their water for more than a month. Three months later large areas of the city still had no water. At the same time, the city was dumping tens of millions of gallons of raw sewage into Lake Pontchartrain every day.

On a sunny November morning, five weeks after Katrina, Kelly Mulholland's rented Ford Explorer bounced across the top of an immense pile of gravel. On one side, brown water rippled through the Industrial Canal. On the other side, workers had already begun to cut through the twisted remains of the heavy steel sheet piling that had once held the levee in place. Beyond that, ten to twenty feet below the surface of the water in the levee, lay the mangled skeleton of the lower ninth ward of New Orleans. At the base of the levee, a massive river barge lay like the carcass of an immense steel whale. The remains of a school bus protruded from beneath the hull, crushed as the boat washed over the levee and beached itself amid the fury of Katrina. As we drove, I tried to imagine the moment that a man-made tsunami some twenty feet high pushed the concrete aside and twisted the thick steel with a horrifying groan before descending on the helpless neighborhood, lifting houses, cars, and trees and tossing them about like children's bathtub toys. Lives, lifetimes, and livelihoods disappeared in an instant.

Given some glitter and spandex, Mulholland could pass for a refugee from World Wrestling Entertainment. With a square jaw, bald head, pale brown goatee, and strapping frame, he might appear menacing were it not for his garrulous good nature. He had come to New Orleans with a team of workers from Portland to help with the recovery effort in the wake of Katrina and wanted to show me the challenge they faced.

As we drove into the ninth ward, past the checkpoint that controlled access to the area, and toward the levee break, the level of destruction built to a crescendo. Fragments of flooded households littered the road. Washing machines and refrigerators lay uselessly rusting on lawns and sidewalks. The twisted remains of an entire child's playset hung from the branches of a tree. A motorboat blocked the road. Another, still on its trailer, rested upside down on a fence.

An "X" in fluorescent orange spray paint with cryptic notes in each corner marked each house like some grim graffiti of disaster. Mulholland explained the markings, "The number on top is the date, on the right is the number of animals, and on the bottom is the number of people inside. A box around the "X" means there is a dead body."

At first most of the Xs had a 0 on the bottom. Mulholland pointed to a small single-story bungalow. "Look," he said, "there's a box." Two blocks later, we saw another. As we approached the levee, we saw more and more. Then, on one corner, four houses, neighbors, all with boxes. As we got closer still, the grid of houses and streets began to break down. The water had lifted the small houses, leaving them twisted at odd angles. Some sat in the middle of the street. Others had been lifted and dropped on top of cars and trucks. Then as we approached the long mound of gravel that had replaced the busted levee, the houses began to disintegrate. Reduced to splinters there were no walls left to spray paint.

Much of the city was still flooded when Mulholland first arrived in New Orleans. The city's drainage pumps normally had the capacity to remove 22.5 million gallons of water from the city every minute, but most of the pumps had been submerged. Many were ruined and others were useless, deprived of electricity. As America focused on the problem of removing the water that had devastated New Orleans for weeks, Mulholland and his team were trying to get water back into the city.

As a critical element in the vast, hidden infrastructure that allows modern cities to exist, water utilities constantly plan and prepare for the disasters that can disrupt or destroy that infrastructure.

Portland has some of the nation's experts in disaster recovery and had sent Mulholland and his team to help. This remarkable act of inter-urban altruism was motivated in part by the realization that no city is immune from disaster. Someday Portland might be looking to the outside world for help. "You never know," said Mulholland, "when a tsunami is going to leave us under twenty feet of water."

The water treatment plant was up and running by the time Mulholland's team arrived, but most of the city did not have safe water. New Orleans drinking water runs through more than sixteen hundred miles of pipe laid out in a complex maze buried just a few feet below ground. Throughout the city magnificent live oaks have wrapped their roots around these pipes like a thousand boa constrictors. As Katrina ripped these ancient trees from the soggy ground and tore houses from their foundations, she left twisted and mangled pipes, most of them still hidden beneath the surface. All around New Orleans, water surged into the street from broken pipes.

With pipes that are often more than one hundred years old, the drinking water distribution system is the dark secret, the hidden weakness of almost every urban water supply. Many sections of these networks lie buried beneath subsequent construction. Many stretches of pipe are so old they have descended into terra incognita, lost to maps and memories. Also hidden underground are improper connections that can contaminate tap water. Old and worn, drinking-water distribution systems always leak.

New Orleans was no different. Even before the storm, its system was riddled with leaks. Many of the pipes were more than one hundred years old. Rust, decay, accidents, and even aggressive tree roots had made holes throughout the buried and neglected system. On an average day before the storm, more than 40 million gallons, almost 40 percent of all the water entering the pipes, seeped through thousands of leaks beneath the city.

Leaks are a fact of life in water distribution systems. It is not uncommon for a third of a city's water supply to disappear through leaks. Now two things made matters much worse. First Katrina had ripped hundreds if not thousands of new holes in the system, many of them major leaks. Whole sections of the system would remain shut off until these were fixed.

The other problem involved the contamination of the system. Water supplies rely on the fact that water leaks *out* of the pipes because of the pressure in the distribution system. If pure water is constantly leaking out of a hole, pathogens are unlikely to enter. Loss of pressure is the enemy. When pressure is low, water (and pathogens) can leak *into* the system. When the storm forced the Carrollton Treatment Plant to shut down, the pressure in the pipes dropped to zero. With the distribution system under more than ten feet of floodwater, contaminated water streamed into new and old leaks all over New Orleans.

For two months Mulholland and his team had been working with local crews, moving methodically from one leak to the next, fixing them, flushing the pipes with chlorine, and moving on. Much of their time was spent simply finding the pipes. The maps for many older areas in the pipe network were lost or never drawn. Even with a map, finding the pipes was a challenge. The storm had generated more solid waste in a few days than New Orleans ordinarily produces in thirty years. Mountains of garbage and debris lined the streets and often covered hydrants, meters, and pipelines. Then there were the refrigerators.

More than 100,000 refrigerators full of food had been abandoned. Many had floated out of houses and lay scattered about. All of them were covered with flies and filled with rotten food. Mulholland recalled that when his teams first arrived in New Orleans the houses reeked from the sewage left by the flood waters and the mold that grew in the damp, but the stench of abandoned refrigerators was overpowering. They could hardly go near the houses until the refrigerators were cleared out.

Even after Mulholland and the team from Portland leaves, the staff of the New Orleans Sewerage and Water Board would have years of work ahead. Three months after the storm, work on the water system hadn't even begun in the lower ninth ward and other severely hit areas of the city. As we drove through an underpass on the way back into the center of New Orleans, the truck splashed through six inches of water. I was a bit surprised to see areas of the city still flooded. "Broken main," said Mulholland, "they still can't find the leak."

Some have suggested that global climate change lies behind the record hurricane season that spawned Katrina. Whether or not this is true, increasing global temperatures will certainly raise sea levels and make coastal cities more vulnerable. The increase in severe storms predicted by most climate scientists will also increase the frequency of flooding, the flow of contaminated runoff into our water supplies, and the frequency with which raw sewage flows into our rivers, lakes, and streams.

Whatever the cause, the winds of Katrina blew much of the water supply for New Orleans and the Gulf Coast back to the nineteenth century. One hundred years ago, New Orleans did not treat its water or its sewage. Two hundred years ago, no city in the world had effective water treatment, and wastewater treatment did not exist. The catastrophe of Katrina reminds us of the past, lays bare the illusions of the present, and tells a cautionary tale of our collective future.

Lack of clean water made New Orleans uninhabitable long after Katrina had passed. Many illnesses could be traced to the hurricane and its aftermath including two cases of cholera, but most of these arose from eating contaminated food or wading through toxic floodwater. The story could easily have been far worse. A massive outbreak of waterborne disease was narrowly avoided only because massive shipments of bottled water reached the city in the days after the crisis and because in the end the refugees from New Orleans found other places to go. Places with clean water. Without the bottled water, without other sources of clean water, the scale of the disaster would have been difficult to imagine. To offer a glimpse of the true potential of water to do harm we need only travel to 1994 and the African city of Goma.

THE WORST PLACE
ON EARTH

As the sun rose over the Democratic Republic of the Congo (DRC), a lone cargo helicopter lumbered through the shadows of the Virunga mountains. Ahead, past the steaming summit of the Nyiragongo volcano, was Lake Kivu, the highest lake in Africa. In the nineteenth century, vacationing colonists had flocked to the shores of this beautiful mountain lake for a respite from the busy work of exploiting the region's natural wealth. In the summer of 1994, its deep blue waters gave rise to the worst outbreak of waterborne disease in human history, an epidemiological perfect storm.

Inside the helicopter Les Roberts, an epidemiologist for the World Health Organization (WHO), peered out at the lush jungle that clung to the mountainside. Beyond the mountains Rwanda stretched out toward the sunrise. Ahead on the ancient lava flow that separated the mountains from the lake, the city of Goma seemed to be under siege, surrounded by the makeshift shelters of hundreds of thousands of refugees.

In the long and bloody conflict between Rwanda's ethnic Hutus and Tutsis, the tide had recently tipped toward the Tutsis. On July 14 a vast river of desperate Hutus had begun to flow into the DRC. Lake Kivu and the Virunga volcanoes formed a natural funnel that squeezed the refugees through a single border crossing and into Goma. As they arrived the winds of the deadly epidemic were already beginning to stir.

Six days later, as Dr. Roberts flew toward Goma, those winds had reached gale force. He had initially set out by car from Kibale, Uganda, but Congolese soldiers had turned him away at the border for lack of a proper visa. Undaunted he returned to Kibale where he learned that the WHO would be sending a helicopter to Goma the next morning. When it took off, he was squeezed in among the aid workers and crates of food. He still did not have a visa, but as they approached their destination, he had a plan.

The roar of the helicopter reduced its occupants to the solitude of thought. Just two months earlier, Dr. Roberts had been in Atlanta, completing his training as an epidemiologist with the Centers for Disease Control (CDC). Now he was flying to the heart of a surreal world ruled by death and disease. He had been working with the WHO to monitor the epidemic and knew the statistics as well as anyone, but they could only hint at the reality on the ground. The scale of the epidemic in Goma and the speed with which it was unfolding stunned even the most seasoned veterans with experience in refugee camps throughout the world.

When the helicopter landed, Dr. Roberts blended in with workers from the aid agency CARE and helped them unload the relief supplies. As he worked, he watched and waited for an opening. With aid shipments and workers from around the world pouring into the airport, the beleaguered customs officials could not begin to keep pace. After a cursory inspection, they had moved on to other arrivals. Dr. Roberts seized the opportunity. He slipped through the cargo area of the airport and in a few moments disappeared into the confusion that was Goma.

A red-haired cork bobbing on a dark-skinned African ocean, Dr. Roberts made his way from the airport to the command and control center of the United Nations High Commission on Refugees (UNHCR) in Goma. Officials there directed him to the medical facilities of Médecins du Monde (MDM) in the Mugunga camp, one of three camps around Goma. As his motorcycle taxi entered the camp and rattled slowly across the sea of human misery, Les Roberts began to grasp the unique horror of Goma. Along the two-mile road from the airport, hundreds of corpses, sometimes stacked two or three high, awaited pickup by workers in the camp.

The dead were not the victims of Tutsi soldiers. A far more ruthless predator had been waiting for the Hutus as they arrived in Goma. Long before the coming of the Rwandans, a microscopic killer had found its own refuge amid the poverty of the DRC. For years it had moved slowly from one Congolese to the next, leaving a trail of disease behind. The sudden appearance of the refugees, exhausted, malnourished, and thirsty from their flight, presented this killer with a rare opportunity. Within days of their arrival, an explosion of cholera had begun to consume the refugees.

On reaching Mugunga Dr. Roberts went directly to the MDM cholera treatment center. After two months of compiling the death toll from the unfolding disaster in Rwanda, he would finally have a chance to intervene. The faceless numbers had not prepared him for what he found. More than a thousand cholera victims were crammed into an open area the size of a baseball diamond. Every five minutes, one of them would die. Desperate aid workers moved among the diseased, trying to help those that could be saved with the meager and entirely inadequate resources at hand.

The treatment for cholera under these conditions is simple and remarkably effective. Cholera is a disease that kills by dehydration. The vast quantities of watery diarrhea produced by its victims can sap more than a liter of fluid an hour from their bodies, often leaving a shriveled corpse within twenty-four hours of the first symptoms. Oral rehydration therapy requires only that the victim drink a mixture of water and salts in sufficient quantities to replace the lost fluids. Before the development of this simple treatment just over thirty years ago, cholera killed between 30 and 60 percent of those with disease. Proper application of oral rehydration can reduce this fatality rate to less than 5 percent.

Oral rehydration requires clean water. One of the organizers of the aid effort, Dr. Jacques de Milliano, president of Médecins Sans Frontières (MSF, or Doctors without Borders), estimated that they needed 3 million gallons to provide for the refugees. On the day that Les Roberts arrived in the camp, fifty thousand gallons of water were delivered to Goma.

In the morning of that day, the MDM cholera treatment center had taken delivery of a bladder containing thirteen hundred gallons of water. Within two hours it was gone. Most had gone not to the cholera victims, but to thirsty refugees armed with guns and machetes. By afternoon aid workers had little more than sentiment to offer the afflicted. Les Roberts watched as cholera drained the life from those around him. As he stood there amid the horror of the epidemic, a feeling of utter helplessness overwhelmed him. In that moment, he realized that he had come to "the worst place on earth."

By the end of the day, more than a third of the people in the treatment center had died. The scene repeated itself at similar centers in this camp and at the other two camps around Goma. Worse yet most of the victims never made it to the rehydration tents. Many days would pass before the situation improved. On July 24 six thousand of the refugees would die in a single day. The disaster was just beginning.

Roberts and the other aid workers who had converged on Goma had prepared themselves for the challenge of delivering medical care for diseased, wounded, and malnourished refugees. What set Goma apart from other refugee operations was the overwhelming task of managing the dead. For weeks on end, refugees hired by the aid agencies drove along the camp's makeshift network of roads, removing corpses and stacking them like cordwood in the backs of pickup trucks. The trucks unloaded their grisly cargo at one end of the camp, where excavating equipment borrowed from the French military labored from dawn to dark, digging endless trenches to receive the steady flow of bodies.

At the treatment center where Dr. Roberts worked, MDM had hired four refugees who spent the entire day removing the dead. A fifth refugee moved from corpse to corpse attending to cholera's most brutal consequence. She did not remove the dead; she collected the living. She lifted hundreds of infants and children from the arms of their dead mothers and arranged for the care of cholera's orphans.

In the crush of desperate and dying refugees, the walls and barriers that define civilization disappeared. The lines that separate one person from another, one space from the next, life from death, faded. Only the tightly packed bodies of the diseased and the dying together with the presence of medical workers and their supplies delineated the location of the cholera treatment center. One morning the physician managing the center assigned Dr. Roberts and another aid worker to establish a perimeter around the center. If it were to serve as more than a collection point for the dead, they would need to define its boundaries. In particular, they needed to protect the tent's water supply. As they set about putting up a fence, they immediately discovered that they could not drive a single post into the ground. Throughout the camp, just below the surface, an impervious layer of solid volcanic rock made digging impossible. Each post required not a posthole, but a pile of rocks to hold it in place.

They labored for hours under the equatorial sun, making their way past the desperate and the dead, trying to create an island of order amid the chaos, piling rocks around one fence post after another. At one point that afternoon, as Dr. Roberts stepped back to stretch the fabric between two posts, the ground beneath him seemed to give. He pulled back reflexively and turned to look down. The dead body he had stepped on stared back at him. Dr. Roberts adjusted the fence line and moved on. Survival in Goma demanded well-defined borders.

In addition to being an epidemiologist, Les Roberts is an environmental engineer. He understood all too well that treating cholera victims, no matter how well it was done, would not solve the problem. The cholera outbreak in Goma was a monster storm beyond anything in recent memory. The center of the storm was Lake Kivu, the area's only adequate source of drinking water.

It is routine practice at refugee camps to drill wells at locations throughout the camp and carefully protect the wellheads from contamination. At Goma the volcanic rock that made setting fence posts so difficult made digging wells impossible. Bringing water into the camps was not feasible. The trucks and roads available in the DRC could not begin to provide the thousands of truckloads of water required by the refugees. Air transport could only bring in a trickle. To the thirsty refugees, the answer was simple. Lake Kivu contained an inexhaustible supply of water.

When the hurricane of weak and hungry refugees collided with the nor'easter of Goma's geography and geology, the superstorm of infectious disease that followed was almost inevitable. With a single water supply, a slight microbial breeze could set the storm in motion. One can only speculate as to the source of that first wind. Perhaps one of the refugees picked up the disease from a source in Goma and then contaminated the lakeshore. Perhaps cholera was hiding in those clear blue waters even before the first refugees arrived.

Once cholera found its way into the crush of refugees that blanketed the vast hillside above Lake Kivu, the storm began to take shape. The frozen lava beneath their feet that had defied efforts to dig postholes and wells made the construction of latrines impossible. Instead, the excrement from hundreds of thousands of people, including tens of thousands of cholera victims, flowed steadily down the hillside and into the lake. As the disease took its toll on the refugees, even their dead bodies began to appear in the lake.

From the first day that they began to arrive, refugees crowded the shores of the lake, filling whatever container they could find. They had no alternative. They could not afford to be concerned about the vast numbers of bacteria seeping into the lake. A strong enough thirst makes any water clean enough to drink.

Under other circumstances, trucks with automatic chlorinators would be used to transport, treat, and distribute the lake water, but they were simply not available in adequate numbers. With no way to purify the water, nothing stood between the disease and its victims.

Dr. Roberts was death's accountant. Each day, he and half a dozen other epidemiologists set out to estimate how many people in the camps were well, how many were sick, and how many had died. Each time he came to Lake Kivu, he had a sense of tremendous futility. He envisioned an effort to enlist refugees to simply add a chlorine solution to buckets of water at Lake Kivu rather than waiting for chlorinator trucks. He went to his superiors at the WHO and asked that he be relieved from his other duties to organize this effort. They insisted that he attend to his primary task, which was to assemble a body count.

Reluctantly he continued with his grim tally as the cholera epidemic ran its course. After cholera had made its way through the camp, shigella, another waterborne killer, moved in. Shigella is a bacteria that shreds the lining of the gut, causing bloody diarrhea and in many cases death. The counting continued. When Roberts and the other epidemiologists from Goma generated the official estimate of deaths, the result was staggering. More than sixty thousand people had died in less than one month, most of them in just two weeks.

Goma was horrific for its scale and suddenness, but little Gomas happen every day all around the world. Tomorrow, between the time you wake up and the time you eat lunch, diarrheal diseases will have killed more people than Hurricane Katrina. In the next ten days, they will kill more people than all the automobile accidents in the United States in an entire year. In less than two months, the count of their victims will surpass the death toll from the Indian Ocean tsunami. Within one year 2.2 million people worldwide will die from diarrheal diseases. The World Health Organization estimates that 88 percent of those deaths derive from unsafe water and inadequate sanitation and hygiene. Children under five will account for 90 percent of the victims.

Around the world 1.1 billion people lack access to improved water sources and 2.6 billion people lack access to improved sanitation. In the words of former U.N. secretary general Kofi Annan, "We shall not finally defeat AIDS, tuberculosis, malaria, or any of the other infectious diseases that plague the developing world until we have also won the battle for safe drinking water, sanitation and basic health care." The problems facing the drinking water supplies of industrialized countries of the world might seem far removed from the desperate struggle for clean water in the developing world. But as we swarm over the planet in ever-increasing numbers with ever-increasing speed, the protection afforded by distance and oceans is an illusion. Disconnection is no longer possible.

On January 29, 1991, just a year before the outbreak in Goma, the General Office of Epidemiology, Ministry of Health (MOH), in Lima, Peru, received reports of an increase in gastroenteritis in an area of the Pacific coast just north of Lima. The MOH responded immediately and within days microbiological samples from the victims were sent to labs in Peru and to labs at the CDC. The results startled public health officials throughout the Americas. For the first time since the nineteenth century, cholera was on the loose in the Western Hemisphere.

One of the laboratories that received samples belonged to Dr. Eugene Rice at the U.S. EPA Drinking Water Research Laboratory in Cincinnati. As the EPA expert on disinfection, his job was to determine the concentration of chlorine needed to purify drinking water in Peru. His first job, however, was not to kill the cholera, but to grow more. Within days billions of bacteria were flourishing on agar plates in the tropical warmth of the culture room.

When Dr. Rice examined the culture plates, he noticed something unusual. As he expected, most of the plates were coated with smooth disks like drops of wax, each one representing a colony containing millions of bacteria. Then he noticed that a few of the colonies were different. Not smooth at all, these colonies had a rough, tortured surface. When he examined these colonies under a microscope, the bacteria from the rough colonies hung together in clumps of tens or even hundreds of cells. By comparison bacteria from the smooth colonies were evenly dispersed. On a hunch he chose to test cells from the two different cultures separately. When he exposed the two groups of cells to chlorine, the results were chilling.

As expected, chlorine killed the bacteria from the smooth colonies rapidly and effectively. Within ten seconds hundreds of thousands of bacteria were reduced to a handful of survivors. Then Dr. Rice exposed cells from the rough or rugose colonies to chlorine. When he examined samples after thirty seconds, thousands of viable bacteria were floating in clumps in the solution. Even after two minutes, these hardy clusters of bacteria were still present.

The rugose strain, it turned out, produced a gooey slime that caused the bacteria to clump together. Chlorine might kill some of the cells on the outer layers of these clusters, but it could not reach into the core. The cells on the outside had sacrificed themselves to protect their bacterial brethren against the chlorine.

In the developed world, the threat posed by chlorine resistance lies more in the possibilities it raises than in any immediate hazard. The cholera outbreak in Peru did ultimately reach the United States, where 102 people fell seriously ill and one died, but these introduced cases never exploded into an epidemic as they did elsewhere. The universal use of water filtration and sewage treatment coupled with the relatively high level of sanitation limited the experience of the United States to a small number of isolated cases. However, the fact that this sort of resistance can occur raises the possibility that cholera or other pathogens could develop the capacity to circumvent the systems we use to treat our water.

When we take technological aim at a microbe, our target is not only moving, but it is also redefining itself to confound our aim and our weapons. When you look at the world from the perspective of a pathogen, human success in beating back most infectious diseases from the industrialized world has created a vast, underexploited ecological niche. A pathogen that can find its way past the detergents, filters, disinfectants, and antibiotics that we throw up in front of it has hit the mother lode. In terms of evolution, this means that a pathogen with the characteristics necessary to get around these barriers will have the greatest chance of reproducing and infecting others. In other words, when we place a barrier in front of a pathogen, we simply redefine the criteria for success.

The emergence of chlorine-resistant pathogens such as cryptosporidium or toxoplasma as agents of waterborne disease provide clear evidence that treatment-resistant organisms can and will emerge. The appearance of a strain of cholera that exhibits a degree of chlorine resistance suggests that even an old and fearsome waterborne nemesis may have the capacity to reinvent itself.

Filters, too, can be defeated. Regulators and treatment plant operators take comfort in the fact that modern filtration plants can remove 99 percent or more of the cryptosporidium flowing into the plant. Clearly reducing the number of oocysts entering the plant by a factor of a hundred or a thousand will dramatically reduce health risks, but it is essential to recognize two things. First, under conditions that are not unusual, millions if not billions of oocysts could flow into a typical treatment plant, and thousands if not millions will flow out to the consumers. Second, the average oocyst flowing out will not be the same as those flowing in. The filter will remove those oocysts most susceptible to removal, that is, the largest oocysts, leaving behind the oocysts that are best suited to passing through filters.

Just as the emergence of antibiotic resistance has put us in a pharmacological arms race with pathogens, so the emergence of cryptosporidium as a significant human pathogen signals the beginning of a new era in the provision of safe drinking water that will challenge us to develop new strategies for source water protection and drinking-water treatment. Perhaps the most appropriate way to view cryptosporidium is as a microbial "proof of concept." Nature can indeed build an organism that we cannot control with chlorine, which, for almost a hundred years, has been the big gun in our water treatment arsenal. Variants of cryptosporidium or other chlorine-resistant pathogens may emerge that are smaller, more infectious, or more virulent. More important, there is no reason to assume that other pathogens will not develop similar capacity to resist disinfection if they have not done so already.

In the world of public health, diseases caused by newly evolved organisms are referred to as emerging infectious diseases. The CDC devotes an entire journal to their study. In 1975, neither cryptosporidium nor *E. coli* O157:H7 would have made its way onto the list of existing human pathogens. Many other pathogens have joined that list and new ones are added every year. Some of them are also waterborne. It is entirely possible, perhaps even likely, that a pathogen with the durability of a cryptosporidium oocyst and the deadly effects of *E. coli* O157:H7 could emerge in the not too distant future.

In the spring of 2003, a previously unknown viral infection was killing people in Hong Kong. The disease came to be known as Severe Acute Respiratory Syndrome (SARS) because the overwhelming and often fatal lung infection struck with such devastating speed. SARS appears to have moved from pigs to humans and became the subject of a massive global public health effort to contain it. For weeks the evening news carried pictures of workers in their germ warfare suits at hospitals in Hong Kong and Toronto as they struggled to comprehend and contain this terrifying new disease. Less widely publicized were the results of studies showing that human feces could spread the disease. The worst local outbreak in the entire epidemic occurred because a man with the disease stayed in an apartment building with poorly vented toilets. In other words one of the deadliest diseases to emerge in recent years showed that it had the potential to become waterborne.

Simple compassion demands that the developed world take action to improve the water supply of the world's poor. Few actions will have a greater impact on global health. Enlightened self-interest makes this call to action even more urgent. It is no coincidence that deadly diseases from SARS to avian flu began in the developing world. Communities in which water and sewage is untreated or inadequately treated and humans live in crowded conditions and in close proximity with animals provide an ideal setting for the emergence of waterborne pathogens. When that happens, distance will provide limited protection.

Two hundred years ago, mere mountain ranges could hold cholera at bay, and when the deadly disease finally escaped from Calcutta, it took more than five years to travel just five thousand miles. Today we are all next-door neighbors. The stunning speed with which avian flu has spread should remind us how easily new and deadly pathogens could reach us in our shrinking world.

16

THE FUTURE OF WATER:
FROM *E. COLI* TO
AL QAEDA

Coleridge had it almost right. For every gallon of water in the world, less than half a cup is fresh. All but one tablespoon of that is locked away in glaciers and the polar icecaps (global warming notwithstanding). Most of the world's liquid freshwater lies buried below the surface as groundwater. If we want to find a drop to drink, we can, of course, follow the diviner's twitching stick and burrow underground, but rainwater, the very essence of life, crawls slowly into the earth. It can take hundreds of years to replenish an aquifer. Growing cities ultimately must either find a source of surface water or stop growing, but they have a minute fraction of the world's water to draw from. From each gallon, less than one drop flows freely on the surface in lakes, streams, and rivers. On that fraction of a drop, the future of water depends.

Humans have always wanted their water clean. Even Hippocrates warned of the risks of foul, stagnant waters. What has changed over time is not the desire for clean water, but the definition of clean. Before Snow and Koch, it was enough that water simply looked, smelled, and tasted clean. Snow showed us that the invisible could kill, and Koch found the invisible bacteria responsible and demonstrated the impor-

tance of removing them. With the discovery of viruses and the knowledge that something that could not be seen even under a microscope
could kill us, the definition of clean changed again. Then, as scientists
began to recognize the risk of toxic chemicals at low concentrations,
the definition changed once more.

In December 2005, the federal government finally issued the rules
on microbial risk and disinfection by-products agreed upon almost a
decade earlier. On top of the reduction in allowable turbidity levels,
the rules include provisions intended to define water supplies at risk
for contamination with cryptosporidium, and require that those utilities test for oocysts. These changes will address some major gaps in the
system and will improve the quality of our water, but the changes are
evolutionary, not revolutionary. Clean has been redefined once again,
but as always the question remains: what is clean enough?

The history of drinking water is a story of disaster and response.
From cholera to cryptosporidium, disasters have forced inquiry and
change. The inquiries have proven contentious and change has often
taken years or decades, but eventually the changes were put in place
to prevent the recurrence of those exact disasters. Cholera is gone from
London. Milwaukee is unlikely to see another major outbreak of cryptosporidiosis. The improvements, however, have almost always looked
backward. The risk we must fear most is the one we have never seen.
In the uncertain future, emerging diseases, changing climates, poorly
understood pollutants, decaying infrastructure, and the dark hand of
terror all threaten us through our drinking water. Are we prepared?
To answer this question, we must consider each element of our water
supply: source water, treatment, distribution, consumers, and management, as well as the ultimate wild card, terrorism.

THE SOURCE

On a cool winter day, fleets of clouds sail in off the Pacific, founder
on the rugged green mountains south of Mount Hood, and spill their

cargo on the pristine wilderness below. Rain drops onto the out-stretched branches of towering Douglas firs and western red cedars and filters down through the lush canopy before falling silently onto a floor of ferns and pine needles. In the Bull of the Woods Wilderness, that moisture runs across the surface or seeps into the soil, past roots and rocks to feed Bull Run Lake, one of the cleanest lakes in the United States. Twenty-five miles to the east, that pristine water flows through pipes beneath the bustling city of Portland, Oregon.

Every drop of Portland's water falls on federal land managed by the U.S. Forest Service. A special agreement established more than a hundred years ago forbids access to the Bull of the Woods Wilderness unless authorized by the Portland Water Bureau. Only with this level of control can a watershed be fully protected. Other cities claim to have protected watersheds, but none of them can match the Bull Run.

On the other side of the continent, on an average day, the sewage treatment plant for Yorktown Heights, New York, sends 1.5 million gallons of treated sewage into the Muscoot River. This fact would be unremarkable but for the ultimate fate of the water in the river. After leaving the Muscoot, the water ambles through a series of reservoirs, descends into an aging network of pipes, and becomes drinking water for the people of New York City. Yorktown Heights, however, is far from the only sewage treatment plant in the immense watersheds that supply New York. Currently 114 communities send their wastewater into New York's "protected" watersheds.

Sewage treatment decreases the number of pathogens in waste-water, but does not eliminate them. New York City depends on dilu-tion and delay to reduce the health risk from the treated sewage. Any pathogens in the water are assumed to disappear in the vast reser-voirs or to die before they can reach the intake pipe. New York City's Department of Environmental Protection is so confident in its ability to protect this watershed that they do not filter the water. In other words, the only active water treatment that stands between the sewage treatment plant of Yorktown Heights and the taps of New York City comes from the pumps that feed chlorine into the water after it leaves the reservoir.

New York is not alone. A handful of other major cities around the country including Boston, Seattle, San Francisco, and Portland, Oregon, do not filter their drinking water. Instead, they rely on watershed protection and chemical disinfection to ensure the purity of their water. Even those who manage the watershed must occasionally admit that the protection is imperfect. As recently as June 2005, after a heavy rain caused a sharp spike in turbidity, New York's health department warned city residents with compromised immune systems to boil their drinking water.

Source water protection is the bedrock of a safe water supply. Although it is possible to purify even the most grossly contaminated supply, total reliance on water treatment means that any failure of the filtration system will open the door for disaster. For those cities that own or maintain statutory control over large drinking water reservoirs and the watersheds that feed them, source water protection is at least plausible. These fortunate water suppliers can face daunting challenges as demonstrated by New York City's "protected" watershed, but they are far better off than others such as St. Louis, Philadelphia, Washington, D.C., or New Orleans, which must contend with whatever the towns and cities upstream choose to dump into the river. Managers of the water supply for these cities can only dream of watershed protection as they stare up the river at water flowing through a complex quagmire of regulatory authorities.

In fact, most Americans drink treated sewage on a regular basis. The lakes and rivers that provide water to most large cities in the United States routinely receive treated and even untreated sewage in proportions far greater than those found in New York's water supply. Heavy rains can send contaminated runoff and raw sewage rushing toward the intake of water supplies downstream. Dry conditions, on the other hand, reduce the amount of water available to dilute treated contaminants flowing into the lake or river. In some cities a major portion of the water they suck into their drinking water treatment plants can be treated sewage during periods when summer droughts or long stretches of icy weather reduce the amount of fresh water flowing into their water supply.

This contaminated water often flows through multiple states and within the purview of dozens if not hundreds of local governments. Without a strong federal hand, uniform protection of interstate waterways is almost impossible. Unfortunately federal regulations would have us believe that there are two different kinds of water: one, the water that receives our waste, and two, the water we drink. The Clean Water Act, designed to prevent the sort of unbridled dumping of waste that turned Lake Erie into an airless puddle and allowed the Cuyahoga River to catch fire, regulates the contaminants flowing into our lakes, rivers, and streams. The Safe Drinking Water Act defines what contaminants are permissible in the water we drink. Each act has spawned separate bureaucracies, separate scientific communities, separate stakeholders, and surprisingly distinct agendas. Efforts to bridge these worlds and develop an integrated approach to our water supply have met with limited success at best.

Watershed protection will always be the most important component of any system to provide safe drinking water. Our watersheds face pressure from all sides. Growing populations increase the demand for water for all purposes. More water consumption means more wastewater, which must find somewhere to go. Larger populations also mean more industry, more agriculture, and more demand for land. All these things translate into greater threats to the purity of the water we rely on for survival.

TREATMENT

Amid the rolling hills of central Massachusetts, two men climbed into a motorboat, drove out into the middle of a large lake toward a flock of seagulls. As they approached, they pulled out a small arsenal of fireworks, took aim at the birds, and opened fire. The barrage of explosions and flames sent the birds squawking into the air. With that flock dispatched, they moved on to look for more birds. After they had finished harassing any bird they could find in the area, they returned to the dock. One might have expected the local police to be waiting at the dock for these troublemakers, but these men have never been charged, despite regular attacks on the birds in the lake. In fact, they get paid for their efforts.

From Swampscott to Hingham to Framingham, the people of the Boston metropolitan area drink the unfiltered water provided by the Massachusetts Water Resources Authority (MWRA). In 1997, after water from its protected watershed failed to meet federal standards for *E. coli* (also known as fecal coliform) for three consecutive years, the EPA ordered the MWRA to build a filtration plant. Having just built a massive sewage treatment plant as part of a costly effort to clean up Boston Harbor, the MWRA was fiscally exhausted. They were reluctant to return to taxpayers for more money, but their permit to provide water without filtration presumed that there was no significant fecal contamination in the water that arrived in Boston from their protected watershed. So the MWRA began preliminary planning for a filtration plant. Then after extensive research, they came upon an alternative plan to avoid filtering their water. They decided to scare the birds.

Birds, as it turned out, were a major source of the *E. coli* in the water supply. In the days before West Nile fever and avian influenza, one might have wondered why a bit of bird poop should threaten us, but the possible risk should now be apparent. After identifying the source of the bacteria, the MWRA concluded that it would be far cheaper to scare birds away from the intake of their water pipes than to build a filtration plant. The program for bird harassment expanded to include hovercraft, propane cannons, egg-smashing expeditions, and elaborate structures to prevent birds from nesting on rocks near the pipe that draws water out of the reservoir. This program, together with a plan to intimidate muskrats and beavers, succeeded in reducing levels of fecal coliform enough to keep the MWRA in compliance with the law.

When the EPA still insisted that they filter, the MWRA decided to fight it in court. One of their main arguments in court was that pipes in Boston's distribution system were so old and corroded that fixing them would be a far better use of their money. Why pour clean water into dirty pipes? The safety of the water supply, they further contended, was a local issue, not a federal issue, and any decision about the best way to protect the water supply should be local. This sounds good, but as the disaster in Walkerton reminds us, local governments do not always have the expertise, resources, and political independence to make the best decisions about drinking-water safety. Boston is not Walkerton, but one can only imagine the state of America's water supplies if Congress had left drinking-water quality as a local issue and had not passed the Safe Drinking Water Act in 1974.

The MWRA won in court, but the victory may have been Pyrrhic. If pouring clean water into dirty pipes was a mistake, is pouring water full of organic matter into clean pipes any better? The MWRA's own estimates put the cost per household for a filtration plant at less than ten cents a day. Is that really too much to pay for an extra measure of confidence in the water supply? Why, one might reasonably ask, didn't Boston want to have the best water possible?

Watershed protection is an essential part of providing safe water, but even the most protected watershed is not immune from contamination. In 1995 in a remote Canadian watershed, mountain lions with toxoplasmosis are believed to have contaminated the drinking water reservoir for Victoria, British Columbia. Toxoplasmosis usually begins with fevers, rashes, or flulike symptoms, but when the organism reaches the brain, it begins to turn the delicate networks into Swiss cheese. At its worst, the disease causes blindness, mental retardation, and even death. The prompt response of the medical and public health authorities in Canada minimized the impact of the outbreak, but not before the disease struck more than one hundred people. Toxoplasma, the single-cell protozoa responsible, forms a chlorine-resistant oocyst similar to cryptosporidium.

Cities with tightly controlled, "protected" watersheds for their source water are in many ways far ahead of cities that depend on multiple-use lakes and rivers for their source water, but watershed protection alone cannot guarantee safe water. Water treatment offers an extra measure of safety for cities with protected watersheds. For cities with unprotected watersheds, purification is the primary line of defense against disaster. In other words, no city can rely on untreated drinking water. So for any water supply the critical question is, how much treatment is enough?

The reduction in allowed turbidity levels and the plans to test for cryptosporidium oocysts in some water supplies specified in the EPA's 2005 rules for surface water treatment helped plug some of the gaps that allowed billions of oocysts into Milwaukee's drinking water. In the name of economic efficiency, however, these rules avoided any call for a sweeping renovation of America's treatment plants. Only the most decrepit or inadequate plants will see significant new construction. There was no call for a significant change in the underlying method for water purification. Instead most of the improvement came from optimizing the operation of the existing plants. Optimizing treatment plant operation is laudable, but why must it take a disaster to convince treatment plant operators to run their plants efficiently?

The 2005 rule changes will certainly reduce the risk of water-borne disease in the United States, but the new EPA standards do not mandate water that is dramatically cleaner than the water that flowed out of the Howard Avenue Treatment Plant in Milwaukee during the spring of 1993 and caused the massive outbreak of cryptosporidiosis.

Utilities that merely toe the line drawn by these rules may not be fully protecting their community.

In fact, some utilities are using self-imposed standards that are far stricter than those issued by the EPA. Milwaukee, for example, now seeks to maintain water with one-tenth the level of turbidity allowed by the EPA. One recent study suggests that these standards may provide an adequate measure of safety.

A research team from the University of California at Berkeley, the CDC, the AWWA, and the EPA recently completed a study in Iowa that involved placing filters in homes served by a utility treating water from the Mississippi River. When the investigators compared rates of disease in homes with filters to a group of similar homes without filters, they found no difference in disease rates.

The study replicates an earlier study in Quebec that attributed 35 percent of cases of gastrointestinal illness to drinking water, but it differs in two important ways. First, unlike the Canadian study, participants in the Iowa study did not know if their water was filtered or not. This was intended to eliminate any error related to those without filters simply reporting more illnesses. Second, the management of the water utility in Iowa was aware the study was under way. If a similar study had been undertaken in Milwaukee during the spring of 1993, perhaps the elevations in turbidity would have been met with greater urgency. Perhaps the outbreak of cryptosporidiosis would never have happened.

Nonetheless, the Iowa study suggests that it may be possible to minimize or even eliminate waterborne transmission of existing pathogens using the best available conventional drinking water treatment operating at maximum efficiency. But generalizing based on the experiences of a single plant over a limited period requires caution. For example, a major flood occurred during the course of the study and its impact on exposure to pathogens may have dwarfed the effects of drinking water. More important, the water produced by the treatment plant in the Iowa study had an average turbidity more than 85 percent below the maximum average turbidity allowed by the EPA's new, stricter rules.

Few treatment plant operators run their plants under the intense scrutiny of federal researchers. As time goes on and the public spotlight disappears, one has to worry that vigilance will fade and water that just meets the standards will become acceptable again. The misadventures of the Koebel brothers should remind us that looking at one exemplary facility is not likely to give us a full picture of the future safety of drinking water in the United States.

The future quality of our water supply depends heavily on our vision for that future. For reasons of cost control, the EPA regulators settled on a scenario that relies on conventional treatment technology. Some utilities have moved beyond simple optimization to offer a far bolder vision for the future of water.

One such alternative sits on a bluff above the Mississippi River in Columbia Heights, Minnesota. There, on September 1, 2005, water began to flow into one of the most advanced water treatment systems in the world and out to the people of Minneapolis. Each day that plant squeezes 70 million gallons of water into 43 million tiny hollow fibers and out through holes two hundred times smaller than a cryptosporidium oocyst.

The opening of the Columbia Heights Water Treatment Plant disappeared in the vast news shadow of Hurricane Katrina, but it marks a sea change in the treatment of drinking water. Using a technology known as ultrafiltration, it meets a standard for water treatment far beyond that set by the EPA. It replaces a plant built in 1913, three years after an outbreak of waterborne typhoid fever killed 185 people in Minneapolis. Most plants around the United States look far more like the plant completed in 1913 than the one that went online in 2005. Other enlightened utilities around the country have concluded in the wake of the outbreak that it is better to move ahead of the EPA than to rely entirely on their standards. From Racine, Wisconsin, to

San Antonio, Texas, utilities have built or are planning to build plants similar to the one on Columbia Heights.

Forward-thinking water suppliers are also reconsidering their relationship with one of their oldest and dearest friends, chlorine. Accumulating evidence about the possible risks from the by-products of chlorination, together with the appearance of pathogens that resist chlorine, have prompted a new look at our reliance on chlorine. A recent study suggesting that the cancer risk may arise from inhalation rather than ingestion of volatile by-products makes this problem even more daunting. Ozone, the most prominent alternative, can inactivate crypto-sporidium oocysts and under most circumstances it appears to produce far fewer toxic by-products than chlorine. Water suppliers in Europe have a hundred years of experience with ozone since its first use in Nice, France, in 1906. Today, thousands of water treatment plants in Europe and Japan rely on ozone, and, somewhat belatedly, many American cities have taken steps to follow suit. Milwaukee now treats its water with ozone. In the summer of 2005, the MWRA flipped the switch to turn on massive new ozone generators to treat Boston's drinking water. These utilities must still add some chlorine to protect the water from contamination as it passes through the water pipes, but in minimal quantities compared with plants that rely on chlorine exclusively. We can expect more utilities to move away from chlorine as a primary method for disinfection as experience with ozone and research on other alternatives increase and improve available options for disinfection.

The treatment plant that supplies water to the EPA Drinking Water Research Laboratory in Cincinnati passes through thick beds of charcoal after trickling through the more conventional sand filters at the city's Miller Treatment Plant on the Ohio River. The charcoal sucks up chemical contaminants like billions of tiny black sponges. Concerns about the impact of a chemical spill on the water supply prompted the installation of the carbon filter, which removes chemical contaminants using the same technology employed by many home water filters.

Conventional water filtration as relied on by most American cities was never intended to remove chemical contaminants from drinking water, and for most utilities chemical removal remains a low priority.

Modern industry produces and releases tens of thousands of different chemicals. Most, if not all, of them find their way at some level into our water supplies. The EPA regulates fewer than one hundred out of tens of thousands of chemicals that can appear in our water supplies, selecting candidates based on their toxicity and prevalence. Regulators focus on single chemicals as they evaluate risk and set standards. Logistics and costs tend to limit risk assessments to a small set of health outcomes, primarily cancer. It is impossible to look at the full range of human diseases when evaluating health risks. As a consequence, we make the implicit assumption that unexamined risks do not exist.

On several counts we may find a need to reexamine our view of chemical risk in general and chemicals in drinking water in particular. An increasing body of evidence indicates that chemicals in the environment may disrupt the subtle chemical communication systems in our bodies in ways never previously considered, with a broad range of possible health effects. Other research suggests that developing embryos and young children may be particularly vulnerable to these effects. Finally, emerging evidence about the combined effect of chemicals in small concentrations, particularly the thousands of unregulated chemicals, raises serious concerns about our entire approach to chemical regulation and may force a reexamination of our relative inattention to chemical contaminants in drinking water.

MULTIPLE BARRIERS

A patient diagnosed with tuberculosis routinely receives a combination of three antibiotics, a treatment regimen known as triple-drug therapy. These drugs are not always essential to kill the organism responsible. Rather they ensure that if some of the tubercle bacilli have developed resistance to one of the drugs they will be eliminated by the other drugs.

Triple-drug therapy is used to prevent the widespread emergence of strains of *Mycobacterium tuberculosis* that are resistant to one or two antibiotics. If we use a single drug, we would soon find that most tubercle bacilli are resistant and the drug would become useless. If we used one drug at a time and waited for resistant strains to emerge before using the next drug, we would eventually find that none of the three drugs would be effective and we would have no protection against one of the worst scourges in human history.

This in many ways is similar to an approach to water treatment that the American Water Works Association refers to as the multiple barrier method, which is comprised of source water protection, coagulation with sedimentation, filtration, and disinfection. A pathogen might find its way past one or even two of these barriers, but it is highly unlikely that large numbers of pathogens can make it through all four of these barriers. Improving individual barriers and using those barriers in combination are essential to providing and maintaining safe drinking water.

The parallels between tuberculosis and waterborne disease grow more chilling when we move to the poorest regions of the world. High infection rates, limited access to care, and the inability to afford multidrug therapy make the explosive emergence of resistant tuberculosis a constant threat that all too often becomes real. Similarly high rates of diarrheal diseases, inadequate or nonexistent sewage treatment, and limited or nonexistent treatment of drinking water, together with the routine mingling of animals and humans, makes the emergence of new waterborne diseases a dangerous possibility. The use of single barriers for water treatment in these circumstances will encourage the emergence of waterborne pathogens that are difficult to remove from drinking water.

The EPA pays close attention to our treatment plants and has a multitude of regulations that apply to the water flowing out of them. Even if those regulations do not require the use of multiple barriers, they implicitly encourage the practice. Those rules, however, have a fundamental limitation: they do not directly regulate the water we drink. Almost all the regulations that apply to drinking water end at the boundaries of the treatment plant. As water crosses the fence line, it enters an obscure subterranean world ruled by rust, slime, and history. That dark and corroded underworld offers uncounted and unseen opportunities for the degradation of our drinking water. Other than a bit of leftover chlorine, nothing stands between a pathogen in the pipes and the glass of lemonade your child is mixing in the kitchen.

THE PIPES

America's drinking water courses through millions of miles of water mains, the unseen vasculature of a modern city. Thirteen hundred miles of those pipes lie buried beneath the streets of Washington, D.C. In the summer of 2004, Jerry Johnson, the man responsible for those pipes and the water inside them, had a plumbing problem.

Over two thousand years ago, the *plumberii* (literally, lead workers) of the Roman Empire began to lay pipes of lead rather than wood or clay. Lead would remain a mainstay of plumbing for the next two millennia. The use of lead pipes and lead solder was not banned in the United States until 1986. In every major city in the United States, one can find thousands of homes with lead pipes connecting them to drinking water mains. Water, with its remarkable ability to dissolve, carries lead to the faucets of those homes. Older cities, such as Washington, D.C., have the most lead. In the first six months of 2004, one in ten homes tested in the city had lead levels above 74 parts per billion (ppb), almost five times the EPA limit of 15 ppb. Research suggests that blood lead levels above 10 ppb may adversely affect the development of a child's brain. Levels above 45 ppb require urgent medical treatment.

Jerry Johnson had had problems meeting the EPA lead limit since it was introduced in 1996. When the EPA ordered that he solve the problem, Johnson and his advisers turned to a method that had been employed by many other utilities in the same situation and they began to add phosphoric acid to their drinking water. If all went well, the phosphoric acid would form a coating on the interior of the lead pipes that would keep the lead from leaching into the water.

Within weeks Jerry Johnson had a new problem. The city's water mains, old and made largely of cast iron, were far from pristine. Thick deposits of iron oxide formed mountain ranges of hardened rust along the interior of the pipes. These deposits narrowed the pipes to a fraction of their original diameter and provided billions of hiding places for bacteria. Even the smoothest pipes are lined with a layer of slime and microorganisms known as the biofilm. In these old pipes, like the pipes in most water utilities, the biofilm teemed with microorganisms. The phosphoric acid loosened the biofilm and flooded the system with bacteria.

Biofilms can harbor a broad range of microorganisms and can protect them from chlorine. This microscopic menagerie routinely includes the pathogen responsible for legionnaire's disease as well as a strain of tuberculosis that can cause severe disease among susceptible individuals, particularly those with lung disease or weakened immune systems. Many bacteria form spores or other dormant forms that are resistant to disinfection and can hide in the biofilm. These include *E. coli* O157:H7, the deadly organism responsible for the Walkerton outbreak. These spores are rarely released in large enough numbers to cause disease outbreaks, but their presence should give us pause.

Although the problem with the water in Washington was solved and no outbreak was detected, the incident hints at some of the problems hidden in the pipes that carry our water. The pipes are riddled with leaks that can be an open door for microbes. The complex connections among those pipes can, under certain conditions, pull contaminated water into the system. The simple fact that these systems have been built and rebuilt by thousands of people over the course of a century or more makes some of these problems inevitable.

In Washington, as in most large cities, the pipes themselves are often more than a hundred years old. Periodically utility workers maintaining these pipe networks even unearth sections of wooden pipe. In 2006, about 20,000 miles of water mains in the United States will need replacement. By 2020 we will need to replace 100,000 miles each year. The EPA has estimated that simply maintaining our water distribution systems in the United States will cost $184 billion in today's dollars over the next twenty years as we replace about 1 million miles of pipes. After that it gets expensive. During the twenty years after 2026, we will need to replace 6 million miles of pipe, at costs in excess of half a trillion dollars.

A peculiar bit of manufacturing history is helping to drive this problem. The first iron pipes were installed in the nineteenth century and have now reached or exceeded their design life of 125 years. About twenty-five years later, manufacturers learned to make a thinner pipe. The thin walls made it less expensive and easier to work with, but reduced the design life to about one hundred years. Another twenty-five years passed before the introduction of a new pipe with still thinner walls. It's design life? Seventy-five years. As if part of some grand, unintelligent design, this pattern of manufacturing innovation has synchronized the decay of all three different types of pipe such that they will all reach the end of their useful life at roughly the same time.

The invisible decay of our water pipes is widely seen within the industry as the Achilles' heel of our system for drinking-water treatment. Generating the political and public support for spending hundreds of billions of dollars to replace rotten pipes will pose a daunting challenge. Water supplies operate largely out of the public view. In the drinking-water industry, public attention almost always implies a problem.

The sheer unsexiness of running dirty water through sand makes generating the political support necessary for vast capital expenditures on drinking water difficult. New treatment plants at least provide the opportunity for a ribbon cutting and a name over the entrance, features capable of generating political support. If water purification lacks political sex appeal, then the rusting cast iron pipes beneath our streets have all the allure of a fungal infection. Replacing them disrupts traffic and costs money. To make matters worse, no one, to my knowledge, has ever expressed interest in having a stretch of buried pipe named in his or her honor.

Although most utilities are moving forward with improvements in their distribution systems, the EPA has projected a major shortfall in spending in this area. According to the National Academy of Science, this shortfall is likely to get worse over the next forty years. The political impetus to tighten drinking water regulations generated by the Milwaukee outbreak has already faded. Let us hope that it does not take another major disaster to generate the public will necessary to fix our leaky pipes.

Even if we could replace all our water mains tomorrow, another sort of problem, a relatively new problem, would remain. Water distribution systems by definition provide large populations over large areas with access to drinking water. Until recently the openness inherent in this system has seemed to have few negative consequences. Recent events have radically changed that perception.

TERRORISM

Everything I have discussed so far in this book focuses on the problem of incidental and accidental contamination of our water supply. Between the nineteenth century and the morning of September 11, 2001, those who are responsible for drinking water safety had maintained the same focus. Then, as the twin towers of the World Trade Center crumbled to dust and horror, so did all our assumptions about safety and risk.

In the wake of the 9/11 attacks, EPA Administrator Christie Todd Whitman told reporters at a White House briefing, "[W]e are actually feeling very comfortable as far as water supplies are concerned, that it would be very difficult to carry out the kind of attack that could result in true health implications to a general population." The executive director of the American Water Works Association rushed to join her, stating, "Most systems have so much water and such effective treatment mechanisms, that anything less than many tankers full of dangerous agents would be diluted and easily neutralized. It is hard to imagine that anyone would have the ability to deliver such quantities effectively and without detection."

One can only hope that this was political posturing intended to divert terrorist aim. As Richard G. Luthy, a professor of civil and environmental engineering at Stanford University and chair of the Water Science and Technology Board of the National Research Council, warned in a presentation to the House Committee on Science:

> I caution you to question very carefully comments I hear from officials that refer to "truck load" quantities of chemicals being necessary to cause harm because of dilution from the large volumes of water being handled. Well, this simply isn't true ... and all the more so if the goal is fear, anxiety, and disruption.

In February 2002 four Moroccan men lay sleeping in an apartment in the Tor Bella Monaca district in the south of Rome. Preparations were almost complete. Carefully sealed containers held four kilos of a cyanide compound. Their trove of false passports and driver's licenses would allow them to move freely and disappear into the population as needed. A map of Rome lay on the kitchen table. Notes and markings on the map highlighted the location of key points in the city's water-distribution system paying particular attention to the location of the U.S. embassy. As the Italian antiterrorist squad burst in that morning, it appeared the men were only waiting for the precise moment to attack.

Although the diligent efforts of the Italian police thwarted this attack before it could happen, an effective attack of the kind these men appear to have had in mind is possible and could have killed hundreds, to devastating effect. Al Qaeda operatives have in fact conducted extensive research on U.S. water supplies and their control systems and indicated that these are potential targets of their attacks.

Much has been made of the risk from the airborne release of weaponized anthrax spores, but there is no more efficient way to deliver biological and chemical agents to every home and workplace than through a water pipe. Whitman and her advisers at the EPA suggested that the nature of our water supplies ensures that any effort to contaminate them would have almost no chance of causing serious harm. This line of thought focuses on the large storage reservoirs that hold vast quantities of water for months or even years. The resulting dilution and degradation would minimize the threat from any introduced pathogen or chemical. Water treatment plants would further reduce the low concentrations of pathogens that did make it into the water supply.

This logic falls apart in the face of the dark brilliance that sends civilian airliners into skyscrapers. The experts have correctly described the strengths in the design of our water supply systems that have eliminated scourges such as cholera and typhoid from the memory of most Americans. Just as the terrorists conceived a plan built on the sliver of metal that could bypass our anti-hijacking measures, they are well aware of the strengths of our water supply systems and perhaps more aware than we are of their weaknesses.

One evening, while I was attending a meeting in Washington, D.C., I shared dinner with a pair of experts from two major water utilities. In the course of the discussion, one of them swore me to secrecy, leaned across the table, and explained how one could contaminate a major portion of an urban water supply with relative ease. On the off chance that diabolical minds have not figured out the details, I will refrain from offering specifics on how a successful attack might be undertaken, but it will not require truckloads of poison. If we are to stop men who take down skyscrapers with box cutters, we must learn to think like them. An attack with the potential to kill hundreds, sicken thousands, and to cause millions if not billions of dollars in economic damage might require nothing more sophisticated than a small group of men with bags of manure.

Our water supplies are vulnerable. The distribution system is the weakest point in the system. The thousands of miles of pipe are difficult to defend and there is nothing more than low levels of chlorine to protect us against contamination. That chlorine will not inactivate toxic chemicals and is ineffective against many pathogens. A recent study even showed that the amount of chlorine in our water mains would have little effect on anthrax spores. Utilities are busily working to harden these systems, but the simple matter of replacing hundred-year-old pipes already overwhelms them. One can only wonder if they are up to protecting this sprawling piece of our essential infrastructure.

With drinking water and other potential targets, we must recognize that all the rules have changed and that ironclad protection against yesterday's threats offers no guarantees of security today. Preparedness demands that we understand what a few determined men can accomplish with the simplest of tools when extremist beliefs ignite within them a profound disregard for their own lives and a fanatical intent to do harm.

THE CONSUMER

In the shadow of Seattle's Space Needle, a blue-green butterfly tattoo flutters on the belly of a teenage girl as if seeking to escape. With one hand she adjusts the controls on her iPod. In the other she holds a bottle of Fijian bottled water. When I ask her why she drinks it, she pulls on a strand of long brown hair before answering, "Like, it just tastes better." I ask if she thinks it's safer than tap water. She takes a final sip, and tosses the empty bottle into the trash. "Totally."

She is not alone. That clattering sound you hear is the drumbeat of the nearly 50 million empty water bottles Americans toss into trashcans and recycling bins every day. Whether for reasons of taste, safety, convenience, or style, Americans like their water in a bottle. Every year, they consume more than 7 billion gallons of bottled water, a number that is rising by 8 percent per year. On almost every count, this is a dubious choice.

The problems begin with the bottles themselves. Their production requires more than two billion pounds of plastic per year, which translates into millions of barrels of oil consumed and a steady release of toxic waste into the environment. The manufacture of a single bottle requires more water than the bottle will ultimately hold. The transport of these bottles over hundreds or even thousands of miles by ship, train, and truck further adds to the disproportionate ecological impact of bottled water.

Even if we set aside concerns about the environmental impact of the bottles, the water inside may not offer the benefits we imagine. Despite the fact that it costs almost a thousand times more than tap water, there is no guarantee that bottled water is safer. Bottled water is less closely regulated than tap water and is not required to meet stricter standards for purity. In fact, a major portion of bottled water in the United States is nothing more than tap water in an expensive bottle. To be sure, many brands of bottled water are superior to tap water and can offer a valuable alternative, particularly when traveling or after a local disaster threatens the water supply. But economically, environmentally, and in many cases even with respect to disease prevention, they fall short as a replacement for piped water.

MANAGEMENT

When Stan Koebel was asked for the cause of the Walkerton outbreak, he said simply, "Complacency." The staunch belief in the adequacy of the status quo and the dismissal if not outright derision of those who challenge the prevailing belief has, from the time of John Snow through the present, set the table at which disaster dined. The drinking-water industry needs a sea change in attitude. There are many forward-thinking people within the drinking-water industry, but overall it resists change.

Both public health and drinking water treatment were born amid the cholera- and typhoid-contaminated waters of the nineteenth century, but in the years since these twin disciplines have grown far apart. They speak far too rarely and when they do they seem to speak different languages. This distance occurs in part as a consequence of the relative success of drinking-water treatment in the industrialized world and the fact that the application of epidemiology, as the fundamental science of public health, to drinking water has almost always focused on major failures of drinking-water treatment. The appearance of public health personnel at a water treatment plant implies a problem.

There are three ways in which the public health community can and should have a renewed and revised role in the maintenance of drinking water safety. First, the public health and drinking water treatment communities must maintain a constant dialogue, even if it at times becomes adversarial. Improved, ongoing, active waterborne disease surveillance could be a key part of this relationship. A proactive surveillance system would not only give the public health community a legitimate, ongoing oversight role, but it could also provide valuable feedback to the drinking-water industry with respect to any undetected problems in the water supply.

Second, the public health community must be allowed full access to the workings of the water companies. Secrecy has never served the drinking-water industry well. Whether the silence covers turbidity spikes in Milwaukee or *E. coli* in the water of Walkerton, public knowledge is the last, best defense against disaster. Since 9/11, a cone of silence and secrecy has descended on the drinking-water industry. Much of the publicly available information about treatment plants and distribution systems has simply disappeared from view. Some of this secrecy is understandable and some of it is even desirable, but when taken to extremes, the instinct to conceal can pose more danger than the terrorists. For example, when the federal government, after intense prodding by the environmental lobby, funded a study of possible sources of contamination upstream from drinking-water intakes, the industry responded by declaring the results of the study secret, since it could reveal the location of the water intakes (locations that are often widely known and even marked on local maps), because terrorists might use the information. This growing secrecy makes the oversight by the public health community even more essential. Of particular concern is the proprietary view of information taken by the private companies that have begun to take over water supplies, particularly in smaller communities.

Finally, the foundation of any forward-looking program is research. Ample funding from the public and private sector to develop innovative treatment technology and to conduct research to quantify known risks, identify emerging risks, and anticipate future threats to our water supply are essential as growing populations, evolving pathogens, and changing industries place new and unanticipated stresses on our water supplies.

We do not consume water. We contaminate it. Growing populations and increasing competition for finite resources of all kinds pose the greatest threat to the future of humanity. Water is the ultimate resource. The world has water in abundance, but accessible, safe water is scarce and will become increasingly so as demand rises and the volume of wastewater increases. In the arid lands of the world, unsustainable water use has become routine and projections of the future built on the simplest of mathematics paint a stark picture of crowded, thirsty cities amid drained aquifers and disappearing rivers.

In 1849, John Snow put forth the radical theory that cholera can spread through microscopic agents in drinking water. He was greeted with derision and disbelief, but that idea redefined the way we look at our drinking water and played a key role in making the modern mega-city viable. Koch and the microbiologists who followed refined his idea and set the stage for the engineers who would in turn create the infrastructure that made safe water possible. Soon the problem of contaminated drinking water seemed to fade.

Over time, the solution disappeared as well. Like much of the critical infrastructure that makes cities possible, our water supply has dropped from view. We simply expect it to work as we focus our energy and attention on other problems. We do so at our peril.

In a changing world, our water supply faces new and unanticipated threats, but the infrastructure resists change. Just as in Snow's day, those who disturb the status quo and point toward flaws in existing systems and the beliefs on which they are based must contend with opposition born of the obstinacy of entrenched ideas and the cost of change. That resistance ultimately arises not from the drinking water industry, but from us. We provide the tax dollars and the political will that makes change possible.

Katrina reminded us that we should never dismiss a threat simply because nothing exactly like it has ever happened. Levees that rest atop the pilings of denial will inevitably fail.

Katrina also showed us what can happen if we view our infrastructure in isolation. Our efforts to manage the Mississippi River led directly to the steady erosion of the Mississippi Delta and the disappearance of the barrier that protected New Orleans from storm surges. Many experts have suggested that this process may lead to a scenario in which New Orleans becomes a fortress against nature, an island served by pumps and surrounded by immense levees, a last barrier against reality.

We must protect the safety of our water, but doing so will require more than fancy treatment plants and new pipes. If we do not protect and preserve our waterways, the most advanced filter will not protect us. If we do not provide pure water and sanitation for the world's poor, their misery and disease will make their way to our doorstep. If we do not take on the difficult, politically treacherous task of developing realistic plans for sustainable water supplies, we will find a world at war over water. If we do not take on the stewardship of our planet with evangelistic fervor, we will accumulate an ecological budget deficit that future generations can never repay.

AFTERWORD

STRATEGIES FOR SAFE WATER:

A MODEST PROPOSAL

Thousands have lived without love, not one without water.

W. H. AUDEN

G iven the present and future threats to the safety of our drinking water, how do we protect ourselves? The following should be part of any strategy to provide full protection of water supplies in the developed world against present and future threats.

1. **Provide political, financial, and technical support for an aggressive campaign to make safe water universally available.**
Unsafe water is a global disaster with devastating consequences for the world's poor, particularly their children. A multinational effort to provide safe water to the developing world will not only give these children a chance and improve the overall health of their families but will also improve the safety of our own drinking water by reducing the chance that new waterborne pathogens will emerge from these compromised supplies.

2. **Aggressively improve source water protection.**
This is the most important element of a safe water supply. Critical

to this effort is tighter integration between agencies regulating surface water contamination and those regulating water treatment. The
federal government can and should play a key role in helping break
down political barriers to this effort, particularly when watersheds
include multiple states.

3. Conserve water and explore innovative ways to recycle water for purposes other than drinking.

Water quantity and quality are intimately related. The most accessible
source for new water supplies is the water we conserve.

4. Move toward advanced technology for water treatment.

We must continue to explore and develop alternatives to the hundred-
year-old technology still employed by most treatment plants in the
United States. At a minimum, plants producing water that poses
health risks and aging plants approaching the end of their design life
should be replaced with more technologically advanced systems with
the capacity to remove both chemical and microbial contaminants,
particularly when source water is compromised. These improvements
should include a vigorous effort to develop and implement alternatives
to chlorination such as ozonation.

5. Mandate the use of multiple barriers.

Multiple barriers are essential to the long-term success of water treatment. They are particularly important as a tool to control emerging
pathogens. Their use must be part of water treatment, not only in the
United States but also in the developing world.

**6. Repair and replace our aging drinking water infrastruc-
ture, particularly the distribution system.**

Repairing and replacing our aging infrastructure for drinking-water
treatment and distribution will require hundreds of billions of dollars. Making this happen will require a concerted campaign of public
education that should begin sooner rather than later. Repairs to the
distribution system should seek to minimize its vulnerability to terrorist attack.

7. Reinvent our water supply to give special status to drinking water.

The vast sums we must spend to replace pipes give us a unique opportunity to rethink the fundamental concepts that define our current drinking water supply. In particular, we should reexamine the notion that all water entering our homes must meet the same standards despite the fact that we drink less than one percent of it. The remainder goes down toilets, into washing machines, onto lawns, or down the drain. As a consequence, improving the quality of every liter of water we drink requires improving the quality of more than fifty gallons of water we don't.

Reconfiguring homes and businesses to conserve water and, where appropriate, use grey water will help, but drinking will always account for a small portion of our water use, and that water should receive a higher level of treatment. In some communities it may be possible to have a separate system for drinking water, but when this is not possible, point of use (POU) filters in our homes and businesses can offer a valuable alternative.

Properly installed and maintained home filters provide an extra measure of protection, yield water that is superior to tap water, and are often safer than bottled water at one tenth the cost, with far less environmental impact. They can eliminate pathogens that our treatment plants fail to remove and can remove chemical contaminants including the by-products of chlorination that most water supplies leave in the water. They can also protect us against contaminants that enter through flaws in our aging distribution systems. Finally, if terrorists choose to attack our water supplies, home filters add an extra measure of protection.

Can we justify the cost of putting a water filter into every home? Can we count on people to maintain those systems? In fact, many homes now have water filters, but they are installed in a haphazard way as a function of the household's income and level of concern about the quality of their municipal drinking water. Furthermore, this has been done independent from the drinking-water utilities. As a consequence, POU filters have never been seen as part of the overall system for providing safe water.

Existing designs for POU filters will need improvement to ensure they are effective and safe and to reduce the waste associated with used filter cartridges, but these changes are entirely feasible.

The involvement of utilities in the selection and installation of POU water filters would have several advantages. First, it could ensure a level of uniformity in the quality and efficacy of the systems used by consumers. The utilities could also ensure that water filters are properly maintained. Second, it could be viewed as an integral part of the overall treatment system. Utilities could have confidence that occasional occurrences of accidental, incidental, or intentional contamination would have little if any consequence. Furthermore, chlorination by-products could be more easily controlled without concern that reduced chlorine would cause an outbreak. Third, the involvement of utilities could ensure that high water quality is not limited to those who can afford to install and maintain a system. By leasing or renting the filters to consumers, just as some utilities rent water heaters, utilities could spread out the cost of pure water for the consumer. Utilities have tended to look askance at home water filters as the unnecessary product of public health paranoia. We should remember that every waterborne outbreak described in this book could have been prevented by the universal use of POU water filters.

8. **Make improving water quality proactive rather than reactive.**

Water suppliers must move from a mode of operation driven by mere adherence to regulation and response to disaster to one in which improvement of water safety is a constant process. Active, ongoing oversight of drinking water by the public health community and well-funded research on water treatment and the risk of waterborne disease are essential to this process.

BIBLIOGRAPHY AND NOTES

A selection of some of the most useful references used in preparing this manuscript is included below. Additional references, images, links, and other information relevant to the book are available at: http://thebluedeath.com.

CHAPTER 1: THE BLUE DEATH

As mentioned in the book, John Snow never married and died relatively young. This may explain why there is no surviving record of his personal life. As a result I had to rely on his published work and the publications of his contemporaries to draw a picture of Snow by inference. A few general references on Snow and on cholera proved particularly helpful.

General Snow References

The best starting point for looking further into John Snow is the wonderful Web site assembled by Dr. Ralph Frerichs, at the UCLA School of Public Health: http://www.ph.ucla.edu/epi/snow.html.

Not just the best Web site on Snow, this is one of the best Web sites I have encountered on any subject. Among its many resources is a digital library of all Snow's publications (unfortunately, a feature not added until after this book was completed).

By far the most complete biography of Snow published to date is that assembled by a group at Michigan State University. It has a trove of information about Snow:

Vinten-Johansen, P., H. Brody, N. Paneth, S. Rachman, and M. Rip *Cholera, Chloroform, and the Science of Medicine: A Life of John Snow.* Oxford: Oxford University Press, 2003.

The book has a recently added Web site with many relevant publications and images at http://www.matrix.msu.edu/~johnsnow/index.php.

Some of the most useful articles on Snow are:

Froggatt, P. "John Snow, Thomas Wakley, and *The Lancet*." *Anaesthesia* 57, no. 7 (July 2002): 667–75.

McLeod, K. S. "Our sense of Snow: the myth of John Snow in medical geography." *Social Science and Medicine* 50, nos. 7–8 (July 2000): 923–35.

Richardson, B. W. "John Snow, M.D., a representative of medical science and art of the Victorian era." *British Journal of Anaesthesia* 24, no. 4 (1952):267–91 (reprinted from *The Asclepiad*, 1866).

Snow, S. J. "John Snow MD (1813–1858). Part I: A Yorkshire childhood and family life." *Journal of Medical Biography* 8, no. 1 (February 2000): 27–31.

———. "John Snow MD (1813–1858). Part II: Becoming a doctor—his medical training and early years of practice." *Journal of Medical Biography* 8, no. 2 (May 2000): 71–77.

———. "Commentary: Sutherland, Snow and water: the transmission of cholera in the nineteenth century." *International Journal of Epidemiology* 31, no. 5 (October 2002):908–11.

I am indebted to Dr. David Zuck for sharing his unpublished biography of Charles Empson with me.

General References on Cholera

For a history of cholera in Europe with a focus on England: Longmate, N. *King Cholera: The Biography of a Disease.* London: Hamish Hamilton, 1966.

For a history of the medical perspective on cholera: Pelling, Margaret. *Cholera, Fever and English Medicine, 1825–1865.* New York: Oxford University Press, 1978.

The book that inspired John Snow to become a vegetarian and to drink distilled water is a bizarre read and an interesting window into what it meant to be a vegetarian in Victorian England: Newton, John Frank. *The Return to Nature; or, A Defence of the Vegetable Regimen: with Some Account of an Experiment Made during the Last Three or Four Years in the Author's Family.* Pt. the 1st. London: Cadell and Davies, 1811.

The cholera outbreak in Sunderland is described in: Ainsworth, William. *Observations on the Pestilential Cholera (Asphyxia Pestilenta) as It Appeared at Sunderland in the Months of November and December, 1831 and on the Measures Taken for Its Prevention and Cure.* London: Messrs. Ebers and Co., 1832; and Haslewood, William. *History and Medical Treatment of Cholera, as It Appeared in Sunderland in 1831, Illustrated by Numerous cases and Dissections. By W. Haslewood and W. Mordey.* London: Longman, Rees, Orme, Brown, Green, & Longman, 1832.

The subsequent outbreak in Newcastle and Gateshead including detailed descriptions of the treatments offered is described in: Greenhow, Thomas Michael. *Cholera, as It Recently Appeared in the Towns of Newcastle and Gateshead; Including Cases Illustrative of Its Physiology and Pathology, with a View to the Establishment of Sound Principles of Practice.* Philadelphia: Carey & Lea, 1832.

CHAPTER 2: SNOW ON CHOLERA

For the debate in the medical community on cholera when Snow first developed his theory on drinking water see:

The London Medical Gazette, vols. 1–48 (nos. 1–1256); December 8, 1827–December 26, 1851.

Royal College of Physicians of London. Cholera Committee. *Report on the Nature and Import of Certain Microscopic Bodies Found in the Intestinal Discharges of Cholera, on the 17th Oct., 1849.* London, 1849.

The *Lancet* also provides an interesting record of the period. Its founding editor was particularly hostile to the ideas of Snow.

Some of Snow's key publications on cholera offer a fascinating window into the evolution of his thinking on the disease:

On the Mode of Communication of Cholera. London: J. Churchill, 1849.

"The cholera at Albion Terrace." *London Medical Gazette* 44 (September 15, 1849): 504–5.

"On the pathology and mode of communication of cholera." *London Medical Gazette* 44 (November 2, 1849):745–52; (November 30, 1849): 923–29.

"On the mode of propagation of cholera." *Medical Times* n.s. 3 (November 29, 1851):559–62; (December 13, 1851):610–12.

"On the prevention of cholera." *Medical Times and Gazette* n.s. 7 (October 8, 1853): 367–69.

"Communication of cholera by Thames water." *Medical Times and Gazette* n.s. 9 (September 2, 1854): 247–48.

"The cholera near Golden Square and at Deptford." *Medical Times and Gazette* n.s. 9 (September 23, 1854): 321–22.

"On the communication of cholera by impure Thames water." *Medical Times and Gazette* n.s. 9 (October 7, 1854): 365–66.

On the Mode of Communication of Cholera. 2d edition, much enlarged. London: J. Churchill, 1855.

"Further remarks on the mode of communication of cholera; including some comments on the recent reports on cholera by the General Board of Health." *Medical Times and Gazette* 11 (1855): 31–35, 84–88.

"Dr. Snow's report" in *Report on the cholera outbreak in the Parish of St. James, Westminster, during the autumn of 1854 presented to the Vestry by the Cholera Inquiry Committee.* London: J. Churchill, 1855, 97–120.

"The mode of propagation of cholera." *Lancet*, February 16, 1856, p. 184.

"Cholera and the water supply in the south districts of London, in 1854." *Journal of Public Health and Sanitary Review* 2 (October 1856): 239–57.

"Cholera, and the water supply in the south districts of London." *British Medical Journal*, October 17, 1857, pp. 864–65.

"Drainage and water supply in connection with the public health." *Medical Times and Gazette* n.s. 16 (February 13, 1858): 161–62; (February 20, 1858): 188–91.

The competing theory of William Budd can be found in: Budd, William. *Malignant Cholera: Its Mode of Propagation and Its Prevention.* London: Churchill, 1849.

CHAPTER 3: ALL SMELL IS DISEASE

Although somewhat dated, the most thorough biography of Chadwick is: Finer, S. E. *The Life and Times of Sir Edwin Chadwick.* New York: Barnes and Noble, 1970.

See also Chadwick, E. *The Sanitary Condition of the Labouring Population of Great Britain*. (1842, reprinted by Edinburgh University Press, 1965.)

Farr's paper on cholera is, Farr, William. "Influence of Elevation on the Fatality of Cholera. *Journal of the Statistical Society of London* 15 (2), 155–83.

CHAPTER 4: THE *EXPERIMENTUM CRUCIS*

Baly was the physician for the notorious Millbank Prison in London. His report for the Royal College was intended to be the definitive work on cholera, but turns on a tortured effort to explain away the extraordinarily high rates of cholera in the prison. Baly, William, and Sir William Withey. *Reports on Epidemic Cholera*. London: J. Churchill, 1854.

Baly's report on London's water supply forced improvements, but the changes came too late to prevent the third wave of cholera to strike London.

Great Britain. General Board of Health. *Report on the Supply of Water to the Metropolis.* London: Clowes, 1850.

This describes the sanitarians' take on cholera and represents an effort to write the science to fit their beliefs.

Great Britain. General Board of Health. *Report{s} to the General Board of Health, on a Preliminary Inquiry into the Sewerage, Drainage, and Supply of Water, and the Sanitary Condition of the Inhabitants* … London, 1849–1856.

For a description of the study of water in the nineteenth century: Hamlin, Christopher A. *Science of Impurity: Water Analysis in 19th Century Britain*. Bristol: Adam Hilger, 1990.

CHAPTER 5: THE DOCTOR, THE PRIEST, AND THE OUTBREAK AT GOLDEN SQUARE

Snow's various reports on cholera are listed above. Most of his thinking on this outbreak and on the comparison of water supplies south of the Thames can be found in his magnum opus:

On the Mode of Communication of Cholera. 2d edition, much enlarged. London: J. Churchill, 1855.

Other key competing reports are:

Great Britain. General Board of Health. Medical Council. *Appendix to Report of the Committee for Scientific Inquiries in Relation to the Cholera-epidemic of 1854 / presented to both Houses of Parliament by Her Majesty's command.* London: Printed by George E. Eyre and William Spottiswoode, 1855.

The Report of the Cholera Inquiry Committee from Saint James Parish:

Westminster, England. St. James (Parish) with John Snow. *Report on the cholera outbreak in the Parish of St. James, Westminster, during the Autumn of 1854. Presented to the Vestry by the Cholera Inquiry Committee, July 1855.* London: Churchill, 1855.

Whitehead, Henry. *The Cholera in Berwick Street. By the Senior Curate of St. Luke's.* London: Hope & Co., 1854.

For a brief biography of Henry Whitehead see: Chave, Sidney P. W. "Henry Whitehead and cholera in Broad Street." *Medical History* 2 (1958):92–108.

CHAPTER 6: THE GREAT STINK

For the story of Napoleon III and the rebuilding of Paris, see: David H. Pinkney, *Napoleon III and the Rebuilding of Paris* Princeton, N.J.: Princeton University Press, 1958.

John Simon's report duplicates Snow's most important study, gives Snow no credit, and suggests that drinking water's role in cholera was his idea.

Simon, John. *Report on the Last Two Cholera-Epidemics of London as Affected by the Consumption of Impure Water; Addressed to the Rt. Hon. the President of the General Board of Health, by the Medical Officer of the Board. Presented to both Houses of Parliament by Command of Her Majesty.* London: Eyre and Spottiswoode, 1856.

For the story of the London sewers and the Great Stink, see: Halliday, Stephen. *The Great Stink of London. Sir Joseph Bazalgette and the Cleansing of the Victorian Metropolis.* Stroud, Gloucestershire: Sutton, 1999.

CHAPTER 7: THE RACE TO CHOLERA

The lives and work of Koch and Pasteur are well documented. Two of the best biographies of these men are:

Brock, Thomas D. *Robert Koch: A Life in Medicine and Bacteriology.* Washington, D.C.: American Society of Microbiology, 1999.

Debre, Patrice. *Louis Pasteur.* Baltimore: Johns Hopkins University Press, 1998.

Koch's writings and research can be found in:

Koch, Robert. *Arbeiten aus dem Kaiserlichen Gesundheitsamte.* Vol. 3. Berlin: Julius Springer Verlag, 1887.

————. "Wasserfiltration und Cholera." *Zeitschrift fur Hygiene und Infectionskrankheiten* 14 (1893):393–426.

Mollers, Benhard. *Robert Koch: Personlichkeit und Lebenswerk 1843–1910.* Hannover: Schmorl und Von Seefeld, 1950.

For the story of Hamburg, see:

Evans, R. J. *Death in Hamburg: Society and Politics in the Cholera Years 1830–1910.* Oxford: Clarendon Press, 1987.

Breyer, H. *Max von Pettenkofer.* Leipzig: Hirzel Verlag, 1980.

CHAPTER 8: THE SCRAMBLE FOR PURE WATER

The story of Chicago's water is drawn primarily from contemporary accounts in the Chicago *Tribune.* Other useful sources include:

http://www.chipublib.org/digital/sewers/sewers.html

http://www.encyclopedia.chicagohistory.org/pages/1324.html

http://www.chicagohistory.org/history/stock.html

Baker, M. N. *The Quest for Pure Water: The History of Water Purification from the Earliest Record to the Twentieth Century.* New York: American Water Works Association, 1949.

Cain, Louis P. *Sanitation Strategy for a Lakefront Metropolis: The Case of Chicago.* De Kalb, Ill.: Northern Illinois University Press, 1978.

The story of the Jersey City Reservoir was drawn primarily from references drawn from the Boonton newspaper.

CHAPTER 9: THE TWO-EDGED SWORD

Much of this chapter draws on my own personal experience. For more on the science, see:

Bove, F., Y. Shim, et al. "Drinking water contaminants and adverse pregnancy outcomes: a review." *Environmental Health Perspectives* 110 suppl. (February 2002): 61–74.

Cantor, K. P., C. F. Lynch, M. E. Hildesheim, M. Dosemeci, J. Lubin, M. Alavanja, and G. Craun. "Drinking water source and chlorination byproducts. I. Risk of bladder cancer." *Epidemiology* 9, no. 1 (January 1998): 21–28.

Hildesheim, M. E., K. P. Cantor, C. F. Lynch, M. Dosemeci, J. Lubin, M. Alavanja, and G. Craun. "Drinking water source and chlorination byproducts. II. Risk of colon and rectal cancers." *Epidemiology* 9, no. 1 (January 1998): 29–35.

King, W. D., and L. D. Marrett. "Case-control study of bladder cancer and chlorination by-products in treated water (Ontario, Canada)." *Cancer Causes and Control* 7, no. 6 (November 1996): 596–604.

King, W. D., L. D. Marrett, and C. G. Woolcott "Case-control study of colon and rectal cancers and chlorination by-products in treated water." *Cancer Epidemiological Biomarkers Preview* 9, no. 8 (August 2000): 813–18.

Koivusalo, M., T. Hakulinen, T. Vartiainen, E. Pukkala, J. J. Jaakkola, and J. Tuomisto. "Drinking water mutagenicity and urinary tract cancers: a population-based case-control study in Finland." *American Journal of Epidemiology* 148, no. 7 (October 1, 1998): 704–12.

Koivusalo, M., E. Pukkala, T. Vartiainen, J. J. Jaakkola, and T. Hakulinen. "Drinking water chlorination and cancer—a historical cohort study in Finland." *Cancer Causes and Control* 8, no. 2 (March 1997): 192–200.

Morris, R. D., A. M. Audet, I. F. Angelillo, T. C. Chalmers, and F. Mosteller. "Chlorination, chlorination by-products and cancer, a meta-analysis." *American Journal of Public Health* 82 (July 1992): 955–63.

Puente D., P. Hartge, E. Greiser, K. P. Cantor, W. D. King, C. A. Gonzalez, et al. "A pooled analysis of bladder cancer case-control studies evaluating smoking in men and women." *Cancer Causes and Control* 17, no. 1 (February 2006): 71–79.

Swan, S. H., K. Waller, B. Hopkins, G. Windham, L. Fenster, C. Schaefer, et al. "A prospective study of spontaneous abortion: relation to amount and source of drinking water consumed in early pregnancy." *Epidemiology* 9, no. 2 (1998): 126–33.

Villanueva C. M., K. P. Cantor, J. O. Grimalt, N. Malats, D. Silverman, A. Tardon, et al. "Bladder cancer and exposure to water disinfection by-products through ingestion, bathing, showering, and swimming in pools." *American Journal of Epidemiology* 165, no. 2 (January 2007): 148–56.

Villanueva, C. M., F. Fernandez, N. Malats, J. O. Grimalt, and M. Kogevinas. "Meta-analysis of studies on individual consumption of chlorinated drinking water and bladder cancer." *Journal of Epidemiological Community Health* 57, no. 3 (March 2003):166–73. Erratum in *Journal of Epidemiological Community Health* 59, no. 1 (January 2005): 87.

Waller, K., S. H. Swan, G. DeLorenze, and B. Hopkins. "Trihalomethanes in drinking water and spontaneous abortion." *Epidemiology* 9, no. 2 (1998): 134–40.

CHAPTERS 10 AND 11: SPRING IN MILWAUKEE and THE HIDDEN SEED

The story of the Milwaukee outbreak draws on extensive interviews with those involved. Lawsuits and lingering disagreements over who should be blamed for the outbreak and who should receive credit for finding its cause have made many of those involved, particularly those who have been interviewed before, extremely reluctant to grant interviews. It took months just to get permission from the mayor's office in Milwaukee to speak with those involved. This is particularly ironic given that the one shining light through the entire experience of the outbreak was then-Mayor John Norquist's insistence on complete openness in providing information to the public.

Ultimately I did talk to most of the people involved, particularly those in the city and state health departments. Several of the key players at the waterworks are now deceased. No one currently working for the waterworks would provide an interview concerning the outbreak. The director of the waterworks specifically refused my request for a tour of the Howard Avenue Treatment Plant.

In addition to interviews, *The Milwaukee Journal* and *The Milwaukee Sentinel* (then two different papers) provided thorough daily coverage of the outbreak. The scientific literature also provides more precise details on the epidemiology of the outbreak. The most salient of these are:

MacKenzie W., N. Hoxie, M. Proctor, M. Gradus, K. Blair, D. Peterson, et al. "A massive outbreak in Milwaukee of cryptosporidium infection transmitted through the public water supply." *The New England Journal of Medicine* 331, no. 3 (1994): 161–67.

Morris, R. D., E. N. Naumova, and J. K. Griffiths. "Did Milwaukee experience waterborne cryptosporidiosis before the large documented outbreak in 1993?" *Epidemiology* 9, no. 3 (1998): 264–70.

Morris, R. D., E. N. Naumova, R. Levin, and R. L. Munasinghe. "Temporal variation in drinking water turbidity and diagnosed gastroenteritis in Milwaukee." *American Journal of Public Health* 86, no. 2 (1996): 237–39.

CHAPTER 12: DRINKING THE MISSISSIPPI

The EPA Web site includes a trove of information on the Mississippi. The best starting place is: http://www.epa.gov/msbasin/.

Payment, P., L. Richardson, J. Siemiatycki, R. Dewar, M. Edwardes, E. Franco. "A randomized trial to evaluate the risk of gastrointestinal disease due to water meeting current biological standards." *American Journal of Public Health* 81 (6): 703–15.

CHAPTER 13: DEATH IN ONTARIO

Extensive government inquiries tell the definitive story of the Walkerton outbreak. These are detailed in a report which, is available at:

http://www.attorneygeneral.jus.gov.on.ca/english/about/pubs/walkerton/.

http://www.walkertoninquiry.com/.

Concurrent news reports in the Toronto *Globe and Mail* provide some additional information. Some of the CBC reporting is still available online at:

http://www.cbc.ca/news/indepth/walkerton/.

http://www.cbc.ca/news/background/walkerton/index.html.

http://archives.cbc.ca/IDD-1-70-1672/disasters_tragedies/walkerton/.

The most complete single source is Perkel, Colin. *Well of Lies*. Toronto: McClelland and Stewart, 2002.

CHAPTER 14: SURVIVING THE STORM

Hurricane Katrina has of course been recorded in exhaustive detail. Contemporary accounts were drawn from *Water World*, the *Washington Post*, *The New York Times*, and the *Times-Picayune*.

CHAPTER 15: THE WORST PLACE ON EARTH

The descriptions of events in Goma relied heavily on interviews with Les Roberts and Ron Waldman along with reports from the WHO and concurrent news reports, particularly the excellent coverage by *The New York Times*.

"Cholera in Goma, July 1994. Bioforce." *Revue d'Epidémiologie et de Santé Publique* 44, no. 4: 358–63.

For the story of the world's water, begin with the WHO site on the subject: http://www.who.int/water_sanitation_health/en/.

Eugene Rice did not respond to repeated attempts to contact him for an interview, so the brief description his work on cholera relied on his paper: Rice, E. W., C. J. Johnson, et al., "Chlorine and survival of 'rugose' Vibrio cholerae." *Lancet* 340, no. 8821 (1992): 740.

And on an interview with John Morris who has also conducted research on chlorine-resistance in cholera. Morris, J. G., Jr., M. B. Sztein, et al., "Vibrio cholerae O1 can assume a chlorine-resistant rugose survival form that is virulent for humans." *Journal of Infectious Diseases* 174, no. 6 (1996): 1364–68.

A few other references on emerging infectious diseases with relevance to drinking water:

Payment, P. "Poor efficacy of residual chlorine disinfectant in drinking water to inactivate waterborne pathogens in distribution systems." *Canadian Journal of Microbiology* 45, no. 8 (1999): 709–15.

Pelletier, P. A., G. C. du Moulin, et al. "Mycobacteria in public water supplies: comparative resistance to chlorine." *Microbiological Sciences* 5, no. 5 (1988): 147–48.

Ridgway, H. F., and B. H. Olson. "Chlorine resistance patterns of bacteria from two drinking water distribution systems." *Applied & Environmental Microbiology* 44, no. 4 (1982): 972–87.

Shaffer, P. T., T. G. Metcalf, et al. "Chlorine resistance of poliovirus isolants recovered from drinking water." *Applied & Environmental Microbiology* 40, no. 6: 1115–21.

Shrivastava, R., R. K. Upreti, et al. "Suboptimal chlorine treatment of drinking water leads to selection of multidrug-resistant Pseudomonas aeruginosa." *Ecotoxicology & Environmental Safety* 58, no. 2 (2004): 277–83.

The SARS story can be found in:

Lau, J. T., H. Tsui, et al. "SARS transmission, risk factors, and prevention in Hong Kong." *Emerging Infectious Diseases* 10, no. 4 (2004): 587–92.

Leung, G. M., A. J. Hedley, et al. "The epidemiology of severe acute respiratory syndrome in the 2003 Hong Kong epidemic: an analysis of all 1755 patients [summary for patients in *Annals of Internal Medicine* 2004 Nov 2; 141 (9):I63; PMID: 15520417]." *Annals of Internal Medicine* 141, no. 9 (2004): 662–73.

Li, Y., S. Duan, et al. "Multi-zone modeling of probable SARS virus transmission by airflow between flats in Block E, Amoy Gardens." *Indoor Air* 15, no. 2 (2005): 96–111.

CHAPTER 16: THE FUTURE OF WATER

This is a brief list. The Web site offers far more with links to or copies of the specific references.

Source Water

http://www.portlandonline.com/water/index.cfm?c=26426
http://www.nyc.gov/html/dep/watershed/html/history.html
http://www.mwra.state.ma.us/

Treatment and Multiple Barriers

http://www.ci.minneapolis.mn.us/water/index.asp

The Pipes

> http://www.dcwasa.com/about/history.cfm
> http://www.nap.edu/openbook/0309096286/html/index.html

Terrorism and Drinking Water

U.S. GAO. Experts' Views on How Federal Funding Can Best Be Spent to Improve Security: Statement of John B. Stephenson, Director, Natural Resources and Environment: GAO-04-1098T:

> http://www.gao.gov/new.items/d041098t.pdf

Hickman, D. C. *A Chemical and Biological warfare threat: USAF Water Systems at Risk: The Counterproliferation Papers.* Future of Warfare Series No. 3. USAF Counterproliferation Center.

ACKNOWLEDGMENTS

Like a river finally reaching the sea, the contents of this book come from many streams. Among those sources, four in particular made this book possible. I am profoundly grateful both to my agent, Nat Sobel, for taking on the first book of an academic and providing advice and encouragement during its conception; and to my first editors, Dan Conaway and Jill Schwartzman, for their belief in me, their patience during the creation of the book, and their wise editorial input. As the book moved into its final reaches, Jeanette Perez joined the project, picked up what had become a double orphan, and carried it to the finish with skill, intelligence, and enthusiasm.

Many friends, colleagues, and mentors helped lay the foundation for this book. During my training and the early stages of my research on water, Tom Chalmers and Fred Mosteller (now both deceased), Bill Hendee, Al Rimm, and Kwang Lee provided support, advice, and wisdom. Research is a team effort, and my research could not have happened without the help of a host of graduate students and research colleagues, including Rajika Munasinghe, Chao Zhang, Ria Chubin, Italo Angelillo, and Melanie Miller. I am also deeply grateful to those in the fields of drinking water and epidemiology who believed in me and stood behind me when others did not; these include Devra Davis, David Ozonoff, Doug Dockery, Ronnie Levin, David Hoel, Tony Miller, and Tony Robbins.

Special thanks must go to Anne Marie Audet, for her work on the meta-analysis described in the book, and to Bruce Kupelnick, a most remarkable man, who not only helped with my work on meta-analysis but also provided input on the early stages of this book. Elena Naumova, a fine statistician and a generous spirit, played a key role in my subsequent work on drinking water and human health. Others

who read the manuscript at various stages and provided valuable input include Kris Morehouse, Charlie Booth, and in particular my sisters Virginia Morris and Tina Raymond. I must also thank the students at Harvard School of Public Health, whose feedback on a series of lectures on drinking water and public health helped shape and inspire this book.

I drew on many libraries and sources to complete this book, but I am particularly grateful to the staff at the National Library of Medicine, Harvard's Countway Medical Library, the Libraries of Gonzaga University, and the University of Washington and the town of Boonton. David Zuck, the John Snow Society, Margaret Pelling, and James Symons also provided key resources and assistance. I am also indebted to Scott Davis and the Department of Epidemiology at the University of Washington for appointing me as a visiting scholar during my work on this book. This book was written in hundreds of places, but I must acknowledge my friends at the Spike Coffeehouse, the Rocket Bakery, and the Verite Coffeehouse for their many fine lattes and friendly conversations (not to mention the Spike's killer ginger peach smoothies).

I also want to thank the people who took the time to provide interviews concerning recent events. Les Roberts, Ron Waldman, and John Morris shared their experiences in Goma and with cholera. The story of Milwaukee relied on the patient input from those who were there, including Tom Taft, Paul Nannis, Steve Gradus, Tom Schlenker, Paul Biedrzycki, Liz Zelazek, Jeff Davis, Sandy Schroederus, Kathy Blair, Don Behm, Eugene Marks, Kate Foss Mullan, Doug Nelson, Steve Hargarten, and Cathy Miller (whose detailed notes taken during the outbreak provided a critical accurate record of events). Kelly Mullholland and Sara Bott provided essential help on my trip to New Orleans.

Tim Ford, Joan Rose, Erik Olson, and Jeff Foran not only provided essential input in the form of interviews, clippings, and feed-

back, but have also been wise and wonderful professional colleagues. Colleagues other than those mentioned above whose thoughts, comments, and witticisms have contributed to this book include Dennis Juranek, Joel Schwartz, Ephraim King, Cynthia Dougherty, Kim Fox, Rosemary Menard, and Steve Estes-Smargiassi. Christiane Möller provided invaluable help in obtaining original German references.

My family has weathered many storms during the course of this book, and I must thank the host of people who helped us through. The fine doctors and nurses of the children's hospitals in Spokane and Seattle saw us through the worst of those storms while many fine women helped our children (Beverly Filzen, Lisa Day, Amy Baker, Kat Edwards, and Sara Graves deserve special thanks). Family and friends also stepped in to help, particularly Klara Burkard, Marjorie Cave, Charlie Cave, Lindley Morton, and Corrine Oishi.

Above all, I must thank my family for their unwavering support. Darwin, Hana, Skyler, and Sage maintained a remarkable spirit through difficult times and were patient all those times I was too busy to play. Astrid, my wonderful wife and challenging first reader, made all of this possible. Without her, this book would not be.